D1158079

Invisible Forces and Powerful Beliefs

invisible well Tanner scientific zones rigorous Masi
Network study develop worked
spirituality Cacioppo represented Nusbaum
Luhrmann members geographical
boundaries disciplinary epistemological Epley
forces Browning sociality health Curlin Hawkley beyond
mind comfort precincts borders approach in innovative
Small Decety Gilpin Semin human Thisted
Berntson

Invisible Forces and Powerful Beliefs

Gravity, Gods, and Minds

The Chicago Social Brain Network

Vice President, Publisher: Tim Moore
Associate Publisher and Director of Marketing: Amy Neidlinger
Acquisitions Editor: Kirk Jensen
Editorial Assistant: Pamela Boland
Operations Manager: Gina Kanouse
Senior Marketing Manager: Julie Phifer
Publicity Manager: Laura Czaja
Assistant Marketing Manager: Megan Colvin
Cover Designer: Chuti Prasertsith
Managing Editor: Kristy Hart
Project Editor: Betsy Harris
Copy Editor: Krista Hansing Editorial Services, Inc.
Proofreader: Apostrophe Editing Services
Senior Indexer: Cheryl Lenser
Compositor: Nonie Ratcliff
Manufacturing Buyer: Dan Uhrig
Word clouds created using www.wordle.net

© 2011 by Pearson Education, Inc.
Publishing as FT Press Science
Upper Saddle River, New Jersey 07458

FT Press offers excellent discounts on this book when ordered in quantity for bulk purchases or special sales. For more information, please contact U.S. Corporate and Government Sales, 1-800-382-3419, corpsales@pearsontechgroup.com. For sales outside the U.S., please contact International Sales at international@pearson.com.

Company and product names mentioned herein are the trademarks or registered trademarks of their respective owners.

Printed in the United States of America

First Printing October 2010

ISBN-10: 0-13-707545-6
ISBN-13: 978-0-13-707545-4

Pearson Education LTD.
Pearson Education Australia PTY, Limited.
Pearson Education Singapore, Pte. Ltd.
Pearson Education North Asia, Ltd.
Pearson Education Canada, Ltd.
Pearson Educación de Mexico, S.A. de C.V.
Pearson Education—Japan
Pearson Education Malaysia, Pte. Ltd.

Library of Congress Cataloging-in-Publication Data
Invisible forces and powerful beliefs : gravity, gods, and minds / the Chicago Social Brain Network.
    p. cm.
Includes index.
    ISBN 978-0-13-707545-4 (hardback : alk. paper) 1. Cognitive neuroscience. 2. Religion and science. 3. Social psychology. I. Chicago Social Brain Network.
QP360.5.I58 2011
612.8'233—dc22
                        2010015248

Contents

Gravity is an invisible force that holds us to the surface of the Earth, yet the fact that gravity is invisible does not place it beyond scientific scrutiny. Similarly, humans are a quintessentially social species whose need for social connection produces invisible forces on our brain, behavior, and biology that are subject to scientific investigation. Among these are forces that compel us to seek trusting and meaningful connections with others and to seek meaning and connection with something bigger than ourselves. The story of these invisible forces speaks to who we are as a species.

The human brain has evolved under the guidance of selfish genes to produce more than a brain that is capable of powerful, isolated information processing operations. The human brain also has evolved with inherent capacities for social cognition, compassion, empathy, bonding, coordination, cooperation, values, morality, and a need for social connection that extends beyond kin and even other individuals.

The dialogue between science and religion, if properly pursued, can usher in a new era of religious humanism in the leading world religions. Their central beliefs and practices largely would remain intact, but their views of nature and their concerns with health and well-being would be refined through their conversations with the sciences. How this model would work is discussed in terms of the relation between love and health in Christian theology—especially the tension between the agape, caritas, and eros models of Christian love.

The status of the body politic and the status

              Most people feel socially connected most of the time. Felt connectedness is typically taken for granted, but the effects of its absence, as experienced in feelings of isolation, demonstrate that our evolutionary heritage as a social species has potent implications for health and well-being.

From relationships to people and groups to

              It has long been recognized that mental states can impact health and well-being, but the causal pathways have only recently begun to be understood. Thoughts, beliefs, and attitudes can have powerful effects on physiological functions, health, and disease. Examples range from superstitious beliefs associated with voodoo, bone pointing, or other black rituals to the more positive states associated with spirituality. This chapter considers these disparate psychological states and how they might translate into physiological effects with real health implications.

              A special case of social interaction concerns two or more individuals engaging in temporally coordinated actions that imply particular timing patterns such as synchrony or rhythmic turn taking, such as applauding in unison or the "wave" that is produced by thousands of individual sports fans in a stadium. A model to explain such synchronized behavior is proposed in terms of the neural processes that are jointly recruited. One of the main implications suggested by this model is that taking part in or being part of a synchronized social interaction gives rise to a qualitative shift in subjective experience due to the difficulty of applying an individual-centered explanation to collectively produced spontaneous co-action.

Language forms the fabric of our social institutions and makes tangible the nature of our relationships. Although the function of language is typically viewed in terms of the information content it provides, some of the social function of language may depend on the way it affects us. The idea of language impact—how language directly affects our emotions and social connections—may be fundamental to the way the social brain functions to connect people.

Specific brain regions in monkeys contain individual brain cells, or neurons, that respond to both observation and execution of identical hand and mouth actions. Brain imaging in humans has demonstrated that our brains have similarly localized regions with similar properties. Localized brain regions respond when goal-directed actions of the hand and mouth are executed and when the same or similar actions are observed. Interestingly, these brain regions in the human also respond to observation and imitation of facial movements, and appear to be sensitive to their emotional content.

Empathy is thought to play a key role in motivating prosocial behavior, guiding our preferences and behavioral responses, and providing the affective and motivational base for moral development. While folk conceptions of empathy view it as the capacity to share, understand, and respond with care to the affective states of others, neuroscience research demonstrates that these components can be dissociated. Empathy is not a unique characteristic of human consciousness, but it is an important adaptive behavior that evolved with the mammalian brain. However, humans are special in the sense that high-level cognitive abilities (language, theory of mind, executive functions) are layered on top of phylogenetically older social and emotional capacities. These higher-level cognitive and social capacities expand the range of behaviors that can be driven by empathy.

Other minds are inherently invisible. Being able to "see" them requires learning about other minds, attending to other minds, and projecting one's own mind onto others. Seeing minds in other agents can mean the difference between treating others as humans and merely regarding them as objects.

The human motivation for social connection extends beyond the boundary of the human in the (often misunderstood) religious language of anthropomorphism. In this chapter, an infamous sermon from colonial America ("Sinners in the Hands of an Angry God") is used to illustrate the way anthropomorphic language works to incorporate human society in a web of ethical obligations that connect to the natural environment and, by imaginative extension, to the universe as a whole.

Becoming a person of faith is not so much about acquiring certain beliefs, but about learning to use one's mind in particular ways. The often intensely private experience of God is built through a profoundly social learning process.

The beliefs that religious individuals hold about the way God operates in human life are potential factors affecting perceived social isolation. This chapter discusses a specific type of such belief that is common in the history of Christian thought: the belief that God is an invisible force of a rather impersonal sort working for the good in everything that happens. The paper argues that this sort of belief has as great or greater potential than belief in God as a personal friend, to give one the sense that one is never alone. However, the conception of God as pervasive can also lead to inattention and disconnection.

Despite the human need for social connection, many individuals are lonely because they are unable to create meaningful social bonds. Interventions designed to reduce loneliness have been nominally successful, suggesting the need for a better understanding of loneliness, social connection, and the obstacles to forming meaningful connections with others.

Science and religion are inextricably intertwined in the practice of medicine. Science has provided modern medicine with extraordinary diagnostic and therapeutic capacities that can be employed to care for patients. Religions may provide a fuller vision for the worthiness of caring for the sick, a framework to guide the application of medical science in that endeavor, and practices that strengthen the human capacity for treating patients as the mindful persons they are.

Invisible forces that connect individuals to society, or to each other, have effects at both ends of the connection. As humans, we are fundamentally individual and fundamentally social. We encompass both the pursuit of rational self-interest of Homo economicus and the pursuit of approval, belonging, and intimacy of Homo socialis, the former grounded in eros, the latter in agape. These forces acting together represent a signature feature of Homo sapiens (the wise ones) and have contributed a record of influence and impact—both positive and negative—that is unmatched in biology.

Preface

We view our past from the here and now, making it seem like contemporary events are a much larger part of our history than they are. Hominids have been estimated to have evolved about seven million years ago, with our species having evolved only within approximately the last 1% of that period. The human brain was sculpted by evolutionary forces over tens of thousands of years, whereas the human achievements we take for granted, such as civilizations, law, and art, have emerged only during the past few thousands of years. A mere 300 years ago, theology and philosophy were the principal disciplinary lenses through which the world was viewed and from which explanations and instruction were sought. Advances in science over the past 300 years have transformed how we think, act, and live. Nearly every aspect of human existence, ranging from agriculture, commerce, and transportation to technology, communication, and medicine, has been transformed by contemporary science. We have no hesitation to accept scientific explanations of physical entities being influenced by invisible forces such as gravity, magnetism, and genes. But when human cognition and behavior are the objects to be explained, deterministic scientific accounts seem less satisfying for many.

For some people, science and modernity are akin to the apple in the Garden of Eden, responsible for our fall from grace. For others, theology and religion represent little more than the stuff of superstition, with no place in an educated society.

About six years ago, we had the opportunity to create a most unusual group of scholars to examine questions about the invisible forces acting on, within, and between human bodies. Superb scholars who individually had made major contributions to their own disciplinary field—fields as divergent as neuroscience and medicine to philosophy and theology— were invited to form an interdisciplinary network of scholars to consider such questions. The development of these discussions even over the first few meetings truly astonished us all. We decided to share what we learned in this book, which represents a different perspective, in which our understanding of human nature is enriched by serious insights and scrutiny that each perspective has to offer. Theology and religion have always relied on unseen forces as the basis for explanations of human behavior and experience. Science has been able to explicate those forces even if along different lines than originally conceived. As we start to consider some of the more complex aspects of human nature, science and theology may be able to work together to shed light on some of these complexities.

We begin this preface and each chapter with a word cloud produced using Wordle, at www.wordle.net. In the case of this preface, the word cloud illustrates key concepts found in this book. In the case of the chapters, the word cloud in each provides a visualization of the key terms and ideas expressed in that chapter. Each chapter, in turn, represents a contribution led by a particular member to the network, but broadened to reflect the interactions of the network on that topic. Perusal of the word clouds across chapters makes the flow of ideas more visible. Together the chapters speak to who we are as a species and the nature of the invisible forces that make us such a unique species. For instance, humans seem to strive for social connections in a variety of ways, from friendships, to identification with groups, to religious affiliations. A major thesis of this book is that we are fundamentally a social species and that this journey is less a march toward isolation and autonomy than it is a march to competence, interdependence, coordination, cooperation, and social resilience. Guiding us through this journey are our social brains, which have evolved to create anything but a blank slate at birth.

We owe a debt of thanks to many for their contributions and support over the years. But we owe special thanks to Barnaby Marsh for approaching us with the idea of forming such a network and for his many contributions to the network, and to the John Templeton Foundation for its support and encouragement to pursue questions, ideas, and conclusions of our science, regardless of where they led.

1 *

Invisible forces operating
on human bodies

We may believe we know why we think, feel, and act as we do, but various forces influence us in ways that are largely invisible to our senses. Gravity is an invisible force that holds us to the surface of the Earth, and magnetism is an invisible force that we use in everyday life. The fact that gravity and magnetism are invisible to us does not place them beyond

* The Chicago Social Brain Network is a group of more than a dozen scholars from the neurosciences, behavioral sciences, social sciences, and humanities who share an interest in who we are as a species, and the role of biological and social factors in the shaping of individuals, institutions, and societies across human history. The scientists and scholars in the Network differ in background, epistemologies, beliefs, and methods. After five years of working together, we found that a common set of themes emerged in our work despite the differences among us. These themes, which provide a different perspective on how we might think about human history, experience, and spirituality, are examined here and explored in more detail in subsequent chapters.

scientific scrutiny. Similarly, a host of forces have emerged over the course of human evolution to influence our thoughts, emotions, and behaviors. Because many of these forces are elemental, we are dealing with an area of human behavior that has also been addressed for centuries by various religions. Among these are forces that compel us to seek trusting and meaningful connections with others and to seek meaning and connection with something larger than ourselves. The story of these invisible forces speaks to who we are and what our potential might be as a species. In short, it is the story of the human mind.

The mind can be thought of as the structure and processes responsible for cognition, emotion, and behavior. It is now widely recognized that many structures and processes of the mind operate outside of awareness, with only the end products reaching awareness, and then only sometimes. But clearly we know a great deal about the mind from what we experience through our senses. It is common sense that we know the shape or color of an object from simply seeing it.

Or do we? It is obvious that the tops of the tables depicted in the top panel of Figure 1.1 differ in size and shape. You may be surprised to learn that your mind is fooling you—the tops of the table are precisely the same size and shape. If you don't believe it, trace and cut a piece of paper the size of one tabletop and then place it over the other. Self-evident truths can sometimes be absolutely false.

The science of the mind is not unique in this regard. As the historian Daniel Boorstin noted[1]:

> Nothing could be more obvious than that the earth is stable and unmoving, and that we are the center of the universe. Modern Western science takes its beginning from the denial of this commonsense axiom…. Common sense, the foundation of everyday life, could no longer serve for the governance of the world. When "scientific" knowledge, the sophisticated product of complicated instruments and subtle calculations, provided unimpeachable truths, things were no longer as they seemed.

And just as the observation that we roam on stable ground led to the incorrect inference that we are the center of the universe, the observation that we look out onto the world and onto others fosters the mistaken notion that the human brain is a solitary, autonomous instrument whose connections with other brains is of no real import.

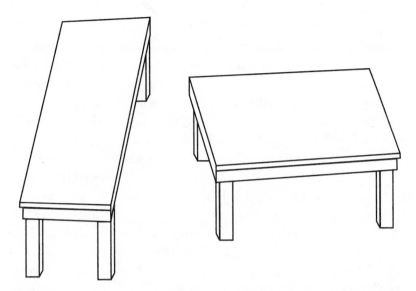

FIGURE 1.1 The two seemingly differently shaped table tops are, in fact, identical in the picture plane. This illusion arises because our visual system provides depth interpretations of the two-dimensional drawing. Table illusion from *Mind Sights: Original Visual Illusions, Ambiguities, and Other Anomalies* by Roger N. Shepard. Copyright © 1990 by Roger N. Shepard. Reprinted by arrangement with Henry Holt and Company, LLC.

The human brain, the organ of the mind, is housed deep within the cranial vault, where it is protected and isolated from others, so it may seem obvious that the brain is a solitary information-processing device that has no special means of connecting with other brains. But we are fundamentally a social species. Faces, expressive displays, and human speech receive preferential processing in neonatal as well as adult brains. When people feel rejected by others, their brains show the same pattern of activation as when they are exposed to a physically painful stimulus. Permit people to cooperate with others, and their brains show the same pattern of activation as when they are given a rich reward such as delicious food or drink. We may not be aware of it, but human evolution has sculpted a human need for social connection, along with neural circuits and hormonal processes that enable and promote communication and connection across brains. As we shall see in the chapters to follow, our sociality is an important part of who we are as a species, and it plays an important, although often invisible, role in the operations of our brain and our biology. Among the questions we examine is whether our social brain also contributes to the ubiquitous human quest for spirituality.

The Chicago Social Brain Network

For hundreds of years, theology and philosophy were the hub disciplines of scholarship, and other fields of inquiry orbited around this dyad and were tightly constrained by it. Over the past three centuries, the sciences have come into their own, displacing theology and philosophy at the center of the academic universe. In so doing, they have produced extraordinary advances in everyday life. People may reminisce about the good old days, but thanks to science and technology, the amount of total income spent on the necessities of food, clothing, and shelter dropped from 80% in 1901 to 50% in 2002–2003. Yet there remains an inchoate sense that something is missing in our lives, something intangible and elusive. Science has improved our material lives, but improvements in material life may not be enough to optimize human well-being.

Can these two very different ways of seeing the world be used synergistically to shed new light on the human mind? To explore this question, in fall 2004, we established an ongoing network of more than a dozen scholars unbounded by disciplinary precincts, geographical borders, or methodological perspectives to set aside antagonisms that had grown up between science and humanities. These Network scholars hail from disciplines as disparate as psychology, neurology, theology, statistics, philosophy, internal medicine, anthropology, and sociology. Each of these scholars was well known in his or her own field and was busy with other obligations, but the opportunity to achieve a deeper, more comprehensive discussion of the human mind made it worth the time and effort required to be part of the Network.

Although various members of the Network interact on a daily or weekly basis, the entire Network convenes twice annually for a four-day retreat to discuss each other's research, critique each other, and learn from one another. Scientific analyses characterized by rigorous experimental designs and data analytic strategies are interlaced with rich philosophical, theological, and historical analyses of the same questions about invisible forces that act on us all. The dialogue between the Network scientists and the scholars from the humanities and theology is bidirectional. For instance, the beliefs and behavior described in the humanities and theology are rich in hypotheses that can now be tested empirically, and the measures and methods of the behavioral sciences and neurosciences now permit rigorous investigation of some of these hypotheses. Each of the Network members brings a unique perspective to the study of the human mind, and the provocative story of the mind that is emerging from the collective efforts of the Network is the subject of this book.

The Network is unconventional in other ways as well. Traditionally, scientists and scholars work together to achieve a common understanding and a consensus position. We quickly learned that we did not need to come to a consensus to benefit tremendously from the dialogue on the capacity and motivation for the ubiquitous human quest for sociality and spirituality. For instance, there is no consensus within the Network on whether there is a God, and we do not seek here to provide the final word on what science and the humanities each have to say to the other about the human mind. Instead, our purpose is to illustrate the possibility and importance of engaging others whose views we may not share in a serious dialogue on such topics. We have learned many lessons as a Network:

1. Some questions about human nature and our social and spiritual aspirations have been asked by humankind for thousands of years. Accordingly, we can gain more from engaging in a collaborative process of thinking about these questions than from demanding simple and immediate answers. We discuss what we see as possible answers to questions about our nature and strivings, but the value in stating these positions is to have clear positions from which to move thinking and research forward. Thus, our purpose in writing this book is to articulate ideas to be shaped and refined, not to provide the final word.

2. One need not agree with a position to perform a deep and thorough analysis of the arguments for and against the position. Objectivity in thought and analysis are keys to reaching a deep understanding of a topic. By taking a position, developing arguments for and against the position, and then taking the opposite position and doing likewise, we develop the capacity to be more dispassionate and powerful thinkers—and gain deeper insight into a topic.

3. One need not reach agreement with someone to learn a great deal from discussions with them or to make significant advances in addressing a complex question. The salve of affirmation can lead us to seek like-minded others and to denigrate and avoid those who disagree with us. Although this may provide temporary comfort, it does little to help address deep divisions or solve problems

that we encounter in an increasingly complex and diverse world. There are inherent tensions between the sciences and the humanities, and these tensions have led to a polarization of views, an "It's my way or the highway" approach toward those holding divergent points of view. The contents of this book illustrate an alternative possibility. The Network is a very interdisciplinary group, and the perspectives captured in the subsequent chapters reflect some of the same tensions that other scientific and religious books have wrestled with—and from which they have not benefited. The tensions reflect deep and enduring differences in the way in which scholars in the humanities, the social sciences, and the sciences think about theory, methods, and evidence. These differences can test one's mettle, but if acknowledged, respected, embraced, and pursued, they result in a richer, more innovative and synergistic collaborative effort. In the case of our Network, this was neither easy nor quick, but it was achieved through a mutual respect and exchange of ideas and a shared conviction regarding the importance of the Network's combination of approaches from the humanities and the sciences. In a sense, our Network is a microcosm of the structure that exists in our society. If these tensions are embraced and used to their full catapultic effect, we can make progress on serious problems, transforming not only how we think about the problem, but also how we think about those who hold different or opposing views.

4. The insights or advances we can achieve need not be our or our opponent's position, or a less than optimal compromise between the two; they can be truly innovative, building on and transcending both initial positions. The specific forms of such creative and transcendent solutions are difficult to articulate in advance, but there is a thought process—characterized by clarity, openness, constructive criticism, and synthesis—that increases the likelihood one will reach such solutions. All of the perspectives discussed in this book have been transformed through this process.

Background

In pursuing the tandem lines of inquiry of science and the humanities, the Network serves as an example of the human capacities and emergent processes that can derive from collective social structures and actions. In the chapters to follow, the Network examines the nature and power of unseen forces, ranging from human coregulation to physiological effects of spiritual beliefs. The exchanges across disciplinary perspectives suggest that the "dominion of the solitary individual" is insufficient to understand the human mind or to optimize human health and well-being. To understand human nature and the human mind, one may need to appreciate human needs and capabilities that have not been given due attention.

Homo sapiens are a social species, which means there are emergent organizations beyond individuals that contribute to the ability of our species to survive, reproduce, and care for our offspring sufficiently long that they, too, survive to reproduce. As a consequence, evolutionary forces have sculpted neural, hormonal, and genetic mechanisms that support these social structures. Among the possible consequences explored in this book are that: 1) people are not the entirely self-interested, short-term-thinking, rational decision makers assumed by the mythical creature *Homo economicus* and 2) some of the amorphous dissatisfaction and chronic diseases that characterize contemporary society may be, in part, the consequence of the denial of the differences between the nature of these two beings. Existing scientific studies of religion have established the pervasiveness of religious beliefs and practices and an association between these beliefs or practices and physical as well as mental health. Religious beliefs and practices have also contributed to failures to heed life-saving medical advice and to the horrendous treatment of others. It will be through the serious investigation of such beliefs and practices, not through their denial, that we may ultimately be able to identify which aspects of these beliefs and practices are beneficial, for what individuals and in what contexts, and through what specific mechanisms.

Recent research has made it patently clear that William James underestimated the faculties of human infants when he suggested that their first sensory experiences were a "blooming, buzzing confusion."[2] But what James's sentiment did capture is the overwhelming complexity and uncertainty that exists in the child's environment, and the inherent difficulty in making sense of that complexity from scratch. Our drive to make meaning is irrepressible—when we do not understand the forces that drive our actions, we invent narratives that make these invisible forces feel more predictable and understandable, even if only in hindsight. But we do not do it alone.

Adults as well as children must explain the uncertainty and ambiguity of natural phenomena (calamities of weather, death, and reproduction) and social phenomena (human agents) to operate effectively. But not all actions are perceived as being equivalent. Forces operating on objects to compel action, as when gravity causes rocks to slide down a mountain, are viewed as external causes. Forces operating on human bodies to produce action, in contrast, are viewed as reflective of purpose, driven not only by external causes but also, more important, by abstract reasons such as goals, aspirations, and destiny. The meaning-making proclivities of humans are so irrepressible that when external forces operate on human bodies to produce a significant impact on humankind, even the causes of the actions of these human bodies tend to be regarded in terms of more abstract purposes and reasons. The anthropomorphic description of hurricanes is a case in point.

Actions of objects have causes, whereas actions of humans have reasons. Invisible forces that operate on humans but that appear to operate independent of human agency have been the subject of religious speculations for centuries. These invisible forces include

- Internal neural and biological forces (such as homeostatic processes and autonomic activity) that exert regulatory forces that are largely hidden from conscious experience or control

- Strong emotions that seem to arise apart from conscious human intention (such as rage, fear, and empathy)

- Phenomena such as dreams or hallucinations that seemingly operate independent from the human will

- Motivations, biases, inclinations, and predilections (such as anthropomorphism, ambiguity avoidance, and preference for simple explanations) whose presence is so universal that, like language, the capacities for their development or expression may have an evolutionary basis

- Individual beliefs (such as the belief that there is a reality outside our head and we are not dreaming

- The belief in human freedom

- The belief in values (such as equality, and so on), attitudes, preferences, goals, or intentions

- Aggregated beliefs that result in social norms, values, religion, culture, and social movements
- Codified forces such as decrees, rules, alliances, and laws

Before the Enlightenment of the eighteenth century, many scholars believed that thought was instantaneous and that action was governed by an indivisible mind separate from the body. If a palpable cause for a person's behavior could not be identified, the Divine or some counterpart constituted a more agreeable explanatory construct than invisible forces acting through scientifically specifiable mechanisms. Unparalleled advances in the sciences have occurred since the dawn of the Enlightenment, including the development of scientific theories about magnetism, gravity, quantum mechanics, and dark matter that depict invisible forces operating with measurable effects on physical bodies. During this same period, serious scientific research on invisible forces acting within, on, and across human bodies was slowed and underfunded in part because the study of the human mind and behavior was regarded by many in the public and in politics as soft and of dubious validity. The result is that many still regard the mind and behavior as best understood in terms of the actions of nonscientific agents, such as a god or gods, and the manifestations of mental illness as the result of a failure of individual will—a denial of the possibility that invisible forces (forces that are tractable scientifically but of which a person is not normally aware) can affect mind and behavior.

One could try to explain away the gap in scientific knowledge about invisible forces by referring to the conception of science and religion as systems of knowledge that are in opposition. This approach is common and evident in a spate of contemporary books that take the position that science and religion represent competing ways of understanding the world, and that science (or religion) is the one and only valid way of understanding human behavior and the world around us.[3, 4, 5, 6, 7, 8] For instance, in *The God Delusion,* Richard Dawkins places specific Judeo–Christian theological doctrines under the scrutiny of science, only to find that none passes scientific muster.

The vast majority of people from all educational backgrounds continue to harbor strong religious beliefs that affect their daily decisions and behavior, with both good and ill effects. These religious belief systems most commonly bump into scientific claims around invisible forces.

When science opens up opportunities to improve the human condition by providing a more complete understanding of the causes of events, their measurable effects, and possible interventions—ranging from valid science education to medical advancements based on stem cell research—these opportunities are often threatened by the application of specific religious beliefs to these endeavors. Scientific research to understand religion and religious belief systems may be a more productive response than broad denouncements by scientists of any who hold such beliefs.

Conversely, when religion opens up opportunities for improving the human condition by questioning the emphasis on short-term self-interests at the expense of the collective, providing a more complete understanding of the human need to attribute meaning to events and their effects, and identifying possible interventions—ranging from the provision of tangible support for individuals in need to the promotion of healthy lifestyles and ethical behavior—scientific research to understand these influences may again be a more productive response than broad denouncements by scientists that such beliefs are irrational. Indeed, the question of whether God exists is of much less scientific interest, and of much more questionable scientific merit (how would one scientifically falsify such a claim?), than the question of the causes, consequences, and underlying mechanisms for the observable human behaviors affected by invisible forces—whether they be physical (gravity), social (groups), or *perceived* spiritual (gods).

Contemporary science explains many of these phenomena but also points to the human capacities and emergent processes that derive from collective social structures and actions and, underlying the emergence of these structures, the human need for meaning-making and connecting to something beyond oneself. The dominant metaphor for the scientific study of the human mind during the latter half of the twentieth century has been the computer—a solitary device with massive information-processing capacities. Computers today are massively interconnected devices with capacities that extend far beyond the resident hardware and software of a solitary computer. The extended capacities made possible by the Internet can be said to be *emergent* because they represent a whole that is greater than the simple sum of the actions possible by the sum of the individual (disconnected) computers that constitute the Internet. The telereceptors (such as eyes and ears) of the human brain have provided wireless broadband interconnectivity to humans for millennia. Just as computers have capacities and processes that are transduced through but extend far beyond the hardware of a single computer, the human brain has evolved to promote social and cultural capacities and processes that are transduced through but that extend far beyond a solitary brain. To understand the full capacity of humans, one needs to appreciate not only the memory and computational power of the brain, but also its capacity for representing, understanding, and connecting with other individuals. That is, one needs to recognize that we have evolved a powerful, meaning-making *social* brain.

Social species, by definition, create structures beyond the individual—structures ranging from dyads and families to institutions and cultures. These emergent structures have evolved hand in hand with neural and hormonal mechanisms to support them because the consequent social behaviors (such as cooperation, empathy, and altruism) helped these organisms survive, reproduce, and care for offspring sufficiently long that they, too, reproduced. From an evolutionary perspective, then, the social context is fundamental in the evolution and development of the human brain.

The observable consequences of these higher organizations have long been apparent, but we are only now beginning to understand their genetic, neural, and biochemical basis and consequences. To fully delve into these complex behaviors, science needs to deal with the invisible forces that shape human life, whether it is in the form of physical, biological, or psychological forces. For instance, anthropomorphism, the irrepressible proclivity to attribute human characteristics onto nonhuman objects to achieve meaning, predictability, and human connection, is beginning to be subjected to productive multilevel scientific analyses. Experimental studies have shown that manipulations that increase feelings of social isolation without the possibility of resolving these feelings through human interaction have the compensatory effect of increasing people's tendency to anthropomorphize, including heightened beliefs in God. This scientific work has implications for understanding claims regarding the success of religious practices, such as solitude, as paths to feeling closer to God. Research on anthropomorphism has now identified developmental, situational, dispositional, and cultural factors that modulate people's tendency to anthropomorphize nonhuman agents, ranging from technological gadgets to animals, to gods, and the neural mechanisms underlying this transconfiguration of nonhuman objects into humanlike agents are beginning to be revealed.

Guided by the insights from these new scientific theories of anthropomorphism, historical analyses may be worthwhile to determine whether concepts of gods have changed across time and cultures such that the god was created in the image of the believer rather than vice versa. For example, in the sixth century B.C., Xenophanes was apparently the first to use the term *anthropomorphism* when describing the similarities between religious agents and their believers, noting that Greek gods invariably had fair skin and blue eyes, whereas African gods invariably had dark skin and dark eyes (joking that cows would surely worship gods that looked strikingly cowlike).[9] In 1841, the theologian Ludwit Feuerbach broached the idea of God as a projection of ourselves. Brain imaging research has confirmed that anthropomorphism is associated with the activation of the same prefrontal areas that are active when people think about themselves or project themselves onto others.[10]

Conclusion

The study of invisible forces also requires a discussion of the method that successful teams use to work together as they cross disciplinary boundaries. Over the past few decades, there has been a demonstrable shift from the individual genius as the source of scientific and scholarly breakthroughs to interdisciplinary teams. This shift in the production of cutting-edge knowledge has been documented in all fields of scholarly activity, ranging from mathematics and theoretical physics to the humanities. This shift has both made possible and been necessitated by a need to understand complex behaviors. Although this project is primarily about the ways that scientists seek to study the impact of invisible forces, it also reflects the methodologies that these researchers use so that their work is not constrained by common knowledge.

The philosophy of science also looks different when dealing with simple causality (one-to-one relations) than with complex causality. Affirmation of the consequent, a logical error in which a given cause for an effect is inferred based on the observation of the effect, does not lead to a scientific error when there is but a single cause for the observed effect. However, as scientific inquiry addresses increasingly complex phenomena, and increasingly complexly determined phenomena, the philosophy of science needs to become more nuanced.

A core challenge is to develop a "science" of identification and aggregation of these invisible forces at different levels. Related research questions include why they exist and measures of robustness. One of our central goals is to demonstrate not only that considerations of these forces matter, but also that they can matter *a lot*.

Questions of value and ethics also could be implicated: Descriptive knowledge, models, awareness of causal relationships, and so on might not be enough to answer some kinds of questions, especially those related to value and purpose, which are the very energies that animate and invigorate real human systems. Economics comes close, with its proxy measure of value based on the distribution of scarce resources and people's varying need for these resources. But this theory comes up short in many instances where other values are at play that are beyond markets, such as in assessing the value of a human life or debating whether all lives are of equal value. It is an especially poor model for helping us understand something as simple as the value of sentimental articles, such as family photographs, which may have little or no market value. Thus, how do we best understand the "sentiments" that are important in the real world?

The members of the Network have worked beyond the boundaries of disciplinary borders, geographical precincts, and epistemological comfort zones to develop a rigorous but innovative approach to the study of the human mind, sociality, spirituality, health, and well-being. The Network members represented in this book are Gary Berntson from Ohio State University, Don Browning from the University of Chicago, John Cacioppo (Network Director) from the University of Chicago, Farr Curlin from the University of Chicago Medical Center, Jean Decety from the University of Chicago, Nick Epley from the University of Chicago Booth School, Clark Gilpin from the University of Chicago, Louise Hawkley from the University of Chicago, Tanya Luhrmann from Stanford University, Chris Masi from the University of Chicago Medical Center, Howard Nusbaum from the University of Chicago, Gün Semin from the University of Utrecht, Steve Small from the University of Chicago Medical Center, Kathryn Tanner from the University of Chicago, and Ron Thisted from the University of Chicago Medical Center. The biography of each member, along with an explanation for the essay each presents, is provided at the beginning of each of their essays on invisible forces.

Endnotes

1. D. J. Boorstin, *The Discoverers* (New York: Random House, 1983), p. 294.

2. W. James, *The Principles of Psychology* (New York: Holt, 1890), p. 442.

3. R. Dawkins, *The God Delusion* (London: Bantam, 2006).

4. D. C. Dennett, *Breaking the Spell: Religion as a Natural Phenomenon* (New York: Penguin, 2006).

5. S. Harris, *The End of Faith: Religion, Terror, and the Future of Reason* (New York: W. W. Norton, 2004).

6. S. Harris, *Letter to a Christian Nation* (New York: Vintage Books, 2006).

7. C. Hitchens, *God Is Not Great: How Religion Poisons Everything* (New York: Hachette Book Group, 2007).

8. D. Mills, *Atheist Universe: The Thinking Person's Answer to Christian Fundamentalism* (Berkeley, Calif.: Ulysses Press, 2004).

9. J. H. Lesher, *Xenophanes: Fragments* (Toronto: University of Toronto Press, 1992).

10. N. Epley, B. A. Converse, A. Delbosc, G. G. Monteleone, and J. T. Cacioppo, "Creating God in One's Own Image," *Proceedings of the National Academy of Sciences*, 106 (2009): 21533-21538.

From selfish genes to social brains

The Chicago Social Brain Network was established to examine how science might inform us about our fundamental human nature, including the apparently irrepressible quest for connection with a higher understanding and organization. Science can describe what religion does in rigorous ways that benefit religion, and religion can serve a meaning-making function that science itself disclaims. This is not to say that science can address the existence of God. Our Network instead focuses on the consequences of believing in such a mind and of seeing into that mind.

In the next chapter, "The social nature of humankind," John Cacioppo, a social neuroscientist, draws on work on evolutionary theory, sociobiology, and evolutionary psychology to examine the implications of the selfish gene hypothesis for *Homo sapiens.* He shows how the notion of the selfish gene has been joined with political theory, consumerism, and economics to produce a dominant modern image of humans, summarized by the phrase "what is best for me is best for the society." Without rejecting the selfish gene view, Cacioppo shows how it evolves in humans into what he calls the "social brain"—a large cerebral cortex and an interconnected limbic lobule that together are sensitive to the complexities of physical and social environments. Central to this complexity is the long period of dependency of the human infant and the interdependencies of adult humans for survival, especially in hostile environments (such as warfare). For the selfish gene to contribute its DNA to the ongoing gene pool, individuals must not only reproduce, but also cooperate with others to ensure that their offspring grow to maturity and reproduce. This leads to natural selection choosing those genes and capacities that contribute to cooperation, reciprocity, attachments, and generosity. During the millennia of human evolution, this process has created the social brain and made humans a unique social animal.

2 *

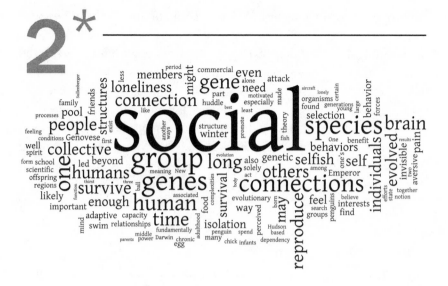

The social nature of humankind

Social species, by definition, are characterized by the formation of structures (such as dyads, families, tribes, and cultures) that extend beyond an individual. Although we may revere the rugged individualist, we are fundamentally a social species. I begin by discussing some of the invisible evolutionary forces that led members of our species to band together to form such structures. I then consider how selfish genes (such as in

* The lead author is John T. Cacioppo, Ph.D., the Tiffany and Margaret Blake Distinguished Service Professor in the Departments of Psychology and Psychiatry, and the Director of the Center for Cognitive and Social Neuroscience at the University of Chicago. He is a cofounder (with Gary Berntson) of the field of social neuroscience, a past president of the Association for Psychological Science, and a recipient of numerous awards, including the National Academy of Sciences Troland Research Award and the American Psychological Association Distinguished Scientific Contribution Award. Cacioppo's research concerns the neural, hormonal, genetic, and behavioral mechanisms underlying the operation

anthropomorphism[1]) might have led to social brains and why the social connections and structures created by humans are especially powerful and flexible. Finally, I describe a nonintuitive way of thinking about the absence of satisfactory social connections (loneliness), mention how and why chronic loneliness can be so harmful, and discuss how our need for social connection motivates us to search for meaning and connections beyond ourselves and other individuals. One implication that is explored here—and covered in more detail in the following chapters—is that genetic and cultural adaptations, not human ignorance, may be fueling the search for meaning and connection with a transcendent entity or being.

Mythic individualism

For at least the past century, we have celebrated the power and intellectual might of the solitary genius. This includes individuals such as Thomas Edison, who brought electrical power to individual households, transforming night into day; Henry Ford, who introduced the mass production of automobiles, changing how we transported and, consequently, how and where we worked and lived; Charles Darwin, who argued that

and maintenance of the emergent structures that characterize social species generally and humans, in particular. He has published more than 400 scientific articles and 16 books, including *Loneliness: Human Nature and the Need for Social Connection* (Norton Books, 2008) and *Handbook of Neuroscience for the Behavioral Sciences* (John Wiley & Sons, 2009). Cacioppo is also the Director of the Chicago Social Brain Network.

Cacioppo has been interested in the similarities and the differences between humans and other species. Human social cognition, emotion, behavior, and executive functioning—that is, our social brain—are especially sophisticated compared to those found in other species. Research in the neurosciences sometimes focuses so much on mechanisms divorced from the social settings and functions they may have evolved to serve in social species such as our own that the generalizations to humans are inaccurate. Animal models permit experimental control and interventions that cannot be carried out in humans, but understanding the implications of this work for the human brain and biology depends on explicit comparison to and knowledge of the rich benefits of human social interaction and feelings of connection. This chapter addresses this gap in our thinking about the genetic, neural, and hormonal processes that constitute our brain and body and, in doing so, provides a different perspective on who we are as a unique biological species.

the difference in mind between humans and other species, great as it is, is one of degree and not of kind; and Albert Einstein, who surmised a relationship between energy and matter, opening a universe of possibilities that previously was virtually unimaginable. As a result, the cultural focus moved from a focus on the social group—the family, neighborhood, or society—toward the autonomous individual.

Forty years ago, the dominant metaphor for the human mind was the digital computer, complete with inputs, filing and memory systems, limited processing resources, and outputs. Evolutionary theory focused on the selfish gene and, by extension, on the individual whose purpose for living was to survive long enough to reproduce. Milton Friedman influenced economic theory and government policies for decades by positing that people, being fundamentally rational, are motivated first and foremost by self-interests, and the adage of "United we stand, divided we fall" was supplanted by the notion that what is best for me is best for the society. Moving from an economy based on manufacturing for the masses to one based on catering to idiosyncratic consumer interests further fueled a focus on the preferences of the autonomous individual.

One can certainly find evidence in humans and other species for the view of life being best understood in terms of self-interest. Sardines, for example, swim in what appears to be synchronized schools until approached by a predator, at which time they dart about so chaotically that they create what seems to be a large, tumultuous ball with a mind of its own. The rule governing the behavior of this dynamic and adaptive collective action can be explained entirely in terms of self-interests. Each fish is driven to swim to the middle, where it is less likely to be eaten by the hungry predator. Sardines are born with the capacity to swim, find food, and avoid predators. If they survive long enough to reproduce, their genes will be part of the gene pool for future generations. That is, if those who are genetically predisposed to swim to the middle are more likely to survive predation, then the genetic predisposition to swim to the middle will become a characteristic of a larger percentage of sardines in future generations.

The sardine ball is an example of a more general phenomenon in which the choices made by members of a group endow the collective with properties that are consistent, predictable, and purposive enough that we can speak of them as "behaviors" of the group, even though the collective actions of the group are not directed by any of the individual members. This phenomenon can be called "emergent" because the properties or behavior of the group are not determined by any individual, but arise from the collective behaviors of the individuals who constitute the group.

According to the National Science Foundation's Tree of Life project, there are between 5 million and 100 million species on Earth, only 2 million of which have been identified thus far. Most of the species identified are either born with the capacity to find sustenance and avoid predation sufficiently well that some survive long enough to reproduce, or they are born in such large numbers that some survive long enough to reproduce. It is the ability of such organisms to reproduce that determines what genes constitute the gene pool for the future generations of that species. These genes, in turn, shape the structure and function of the organisms that constitute a species. This reasoning led George Williams to suggest a half-century ago that traits (that is, behavioral tendencies) that benefit the group at the expense of the individual would evolve only if the process of group selection was great enough to overcome selection within groups.[2] He further suggested that group selection is nearly always weak, so group-related adaptations do not exist.[3] Richard Dawkins[1] popularized the notion that traits that evolve are adaptive at the gene level through his use of the metaphor of the selfish gene. Genes serve their own selfish interests, in the sense that whatever contributions are made by a gene, or set of genes, to an organism's structure and function would be passed on to the next generation only if the gene made its way to the gene pool. Survival of the fittest now had a biological basis.

United we stand, divided we fall

Charles Darwin did not know that genes were the mechanism through which structures and behaviors evolved, but an important component of his original theory was the notion of survival of the fittest. Darwin was also puzzled by the observation that many individual organisms made themselves less fit so that the group might survive. Subsequent generations of evolutionary biologists realized that even though genes might act as if selfish, the *vehicle* responsible for the transport of these genes to the gene pool occasionally extended beyond the individual or parent to kin and even to unrelated members of groups. More specifically, in some cases the superorganismal structures formed by social organisms represent naturally selected levels of organization above individual organisms.[4]

Consider the example of the Emperor penguin *(Aptenodytes forsteri)*. Emperor penguins typically reside near their food source of squid, shoaling fish, and small crustaceans, but they gather into breeding colonies (rookeries) up to 60 miles inland in April and May during the Antarctic winter. They search for their mate from the previous year and go through a courtship ritual before mating. The female lays only one egg in May or June, which coincides with the start of the bitter Antarctica winter. The Emperor penguins are thought to have developed this unusual winter breeding behavior to permit the chick to grow to independence the following summer, when food is plentiful. Ensuring that the chick survives that long is a collective enterprise—the vehicle responsible for the chick surviving long enough such that it, too, can reproduce is not solely the mother or the father, but also the huddle.

The birthing of the egg leaves the mother depleted, so she must return to the seaside to feed while the father assumes responsibility for the incubation of the egg during the winter. An egg from an Emperor penguin will quickly freeze if left exposed to the harsh winter conditions of Antarctica, so the transition of the egg from beneath the warmth and safety of the mother to atop the feet and under a fold of feathered abdominal skin of the father requires a bit of coordination on the parents' part. Even this act is insufficient for the genes of this pair to find their way into the gene pool. The conditions of the Antarctic winters are among the least hospitable on Earth, with winter temperatures dipping below –60° Celsius and winds reaching 120 mph. To protect themselves from the wind and cold, the male penguins huddle together, spending much of their time sleeping to conserve energy. In this harsh environment, survival of the chicks depends on the shared warmth and protection of the huddle, not the individual. The group as a whole is more likely to survive if each penguin and chick shares in the warmth and protection of the collective structure, which means selective pressures exist to promote cooperation to maintain the integrity of the huddle. More generally, for species born to a period of utter dependency, the genes that find their way into the gene pool are not defined solely or even mostly by likelihood that an organism will reproduce, but by the likelihood that the offspring of the parent will live long enough to reproduce. As in the case of the Emperor penguins, one consequence is that selfish genes evolved through individual-level selection processes to promote social preferences and group processes, including reciprocal social behaviors, that can extend beyond kin relationships.[5] Examples of such selection processes in humans exist as well.[6, 7]

The environmental challenges facing Emperor penguins, as daunting as they are, pale by comparison to the complexities facing the human species. Indeed, the social brain hypothesis posits that the social complexities and demands of primate species contributed to the rapid increase in neocortical (the outer layer of the brain) connectivity and intelligence.[8] Warfare among ancestral hunter-gatherers appears to have contributed to group selection for human social behaviors, especially altruistic behaviors.[5] As Darwin noted:

> A tribe including many members who, from possessing a high degree of the spirit of patriotism, fidelity, obedience, courage, and sympathy, were always ready to aid one another, and to sacrifice themselves for the common good would be victorious over most other tribes; and this would be natural selection.

Moreover, deducing better ways to find food, avoid perils, and navigate territories has adaptive value for large mammals, but the complexities of these ecological demands are no match for the complexities of social living (especially in hostile between-group social environments), which include the following: recognizing in-group and out-group members; learning by social observation; recognizing the shifting status of friends and foes; anticipating and coordinating efforts between two or more individuals; using language to communicate, reason, teach, and deceive others; orchestrating relationships ranging from pair bonds and families to friends, bands, and coalitions; navigating complex social hierarchies, social norms, and cultural developments; subjugating self-interests to the interests of the pair bond or social group in exchange for the possibility of long-term benefits for oneself or one's group; recruiting support to sanction individuals who violate group norms; and doing all this across time frames that stretch from the distant past to multiple possible futures).[9] Consistent with this hypothesis, measures of sociality in troops of baboons have been found to be highly correlated with infant survival, and cross-species comparisons have shown that the evolution of large and metabolically expensive brains is more closely associated with social than ecological complexity.[9]

Our survival depends on our connection with others. Born to the most prolonged period of utter dependency of any animal, human infants must instantly engage their parents in protective behavior, and the parents must care enough about their offspring to nurture and protect them. If infants do not elicit nurturance and protection from caregivers, or if caregivers are not motivated to provide such care over an extended period of time, then the infants will perish along with the genetic legacy of the parents.[10] Our developmental dependency mirrors our evolutionary heritage. Hunter-gatherers did not have the benefit of natural weaponry, armor, strength, flight, stealth, or speed relative to many other species. Human survival depended on *collective* abilities, not on individual might.

Selfish genes, social brains

The gene is obligatorily selfish, not the human brain. Genes that promote behaviors that increase the odds of the genes surviving are perpetuated. One implication of this simple insight is that evolution can be viewed as the competition between genes using individuals and social structures as their temporary vehicles. The genetic constitution of *Homo sapiens* in the long run derives not solely from the reproductive success of individuals, but also from the success of their children to reproduce. Hunter-gatherers who did not form social connections and who did not feel a compulsion to return to share their food or defense with their offspring may have been more likely to survive to procreate again, but given the long period of dependency of human infants, their offspring may have been less likely to survive to procreate. The result is selection that strongly favors the ability to process information that could contribute to the formation and maintenance of social capacities and connections— that is, a social brain. These social capacities evolved hand in hand with genetic, neural, and hormonal mechanisms to support them because the resulting social behaviors helped humans survive, reproduce, and care for offspring sufficiently long that they, too, survived to reproduce.[11, 12, 13] Relative to other animals, the striking development of, and increased connectivity within, the human cerebral cortex, especially the frontal and temporal regions, are among the key evolutionary developments in this regard. The cerebral cortex is a mantle of between 2.6 billion and 16 billion neurons, with each neuron receiving 10,000 to 100,000 synapses in their dendritic trees.[14] The expansion of the frontal regions in the human brain contributes to the human capacities for reasoning, planning, performing mental simulations, developing a theory of mind, and thinking about self and others. The temporal regions of the brain, in turn, are involved with aspects of social perception, memory, and communication. The means for guiding behavior through the environment emerged prior to neocortical expansion. The evolutionarily older systems also play a role in human information processing and behavior, albeit in a more rigid and stereotyped fashion. The intricately interconnected neocortical regions of the frontal lobes are involved in self-control, which permits the modulation of these older systems and the overriding of organismal hedonistic impulses for the benefit of others.[15]

Evidence across human history provides overwhelming support for the supposition that humans are fundamentally social creatures.[13] Even in contemporary times in which autonomy is revered, the average person has been estimated to spend nearly 80% of waking hours in the company of others, most of which is spent in small talk with known individuals.[16] These estimates have been supported in more detailed assessment using the day-reconstruction method to determine how people spend their time and how they experience events in their lives on a daily basis.[17] The results of these daily assessments indicate that people spend only 3.4 hours, or approximately 20% of their waking hours, alone. The time spent with friends, relatives, spouse, children, clients, and coworkers is rated on average as more inherently rewarding than the time spent alone.[17]

Respondents indicate that their most enjoyable activities are engaging in intimate relations and socializing—activities that promote bonding and high-quality relationships—whereas their least enjoyable activities are commuting and working. These results are consistent with survey data. When asked "What is necessary for happiness?" the majority of respondents rate "relationships with family and friends" as most important,[18] although we certainly do not always act like this is most important. It is surprisingly easy to overlook the evident.

Noticing the unusual, overlooking the obvious

On January 15, 2009, US Airways Flight 1549 departed from New York's LaGuardia Airport for Charlotte, North Carolina, when it struck a flock of geese during takeoff. Both engines were disabled, and the heavy aircraft quickly lost the power it needed to stay aloft. Capt. Chesley Sullenberger somehow managed a controlled descent into the Hudson River. The media dubbed the ditching of the plane and the survival of all 155 passengers and crew the "Miracle on the Hudson," and Sullenberger was duly heralded as a hero. The ability to control the descent of an 84-ton plane without engine thrust is not something with which humans are naturally endowed. Sullenberger was not a novice, of course. He was a U.S. Air Force Academy graduate who flew F-4 fighter planes while in the Air Force, amassed 40 years of flight experience, and held a commercial glider license and glider instructor rating. As remarkable as his achievement was relative to what one might normally expect in this situation, however, Sullenberger's efforts were not sufficient for the Miracle on the Hudson to be achieved.

When Flight 1549 came to a stop in the frigid Hudson River, the passengers and crew scrambled to the wings and inflatable slides of their slowly sinking aircraft. Local commercial vessels from the New York Waterway and Circle Line fleets responded almost immediately, with the first vessels reaching the plane within four minutes. The crews of the various vessels worked together to rescue the passengers and crew of Flight 1549, and various volunteers and agencies offered medical assistance. These rescue efforts were not motivated by personal or commercial self-interests, and none of the commercial vessel captains was lauded as a hero. Their efforts received less attention because their actions were precisely what we expect of one another.

It is the unusual, not the commonplace, that is noticed. On March 13, 1964, Kitty Genovese parked near her home in Kew Gardens, New York, and proceeded to her residence in a small apartment complex. Winston Moseley, a business machine operator who later confessed that his motive was simply to kill a woman, overtook Genovese and stabbed her twice in the back. Genovese screamed, "Oh my God, he stabbed me! Help me!"—a call that was heard by neighbors. When one neighbor shouted at the attacker, "Leave that girl alone," Moseley ran away.

Genovese, who was wounded and bleeding, moved toward the apartment building slowly and alone. Moseley returned approximately ten minutes later and searched for Genovese. Finding her nearly unconscious in a hallway of the building, he continued his knife attack on her and sexually assaulted her. The entire attack unfolded over about half an hour, yet no one responded. The first clear call for help to the police did not occur until minutes following the final attack, and Genovese died in an ambulance en route to the hospital. The number of people who were aware of some aspect of the attack was estimated to be from one dozen to more than three dozen. One unidentified neighbor who saw part of the attack was quoted in *The New York Times* article as saying, "I didn't want to get involved."[19] The notion that people might not go to the aid of another, even a stranger, in dire need led to public outrage. Decades of research led to the conclusion that the ambiguity of the situation and the diffusion of responsibility were contributing factors.

These two news stories illustrate, in very different ways, how invisible forces sculpted by evolution and cultivated by the environment act on our species. When commercial boat captains act against their own financial interests to rescue others on a sinking aircraft, we think nothing of it because we believe it is what any individual in the same situation would naturally do. When observers of a brutal attack do nothing to aid the victim, we are horrified because we believe it goes against who we are as a species. Humans are not motivated solely by self-interests; we work together and help one another when in need. We survive and prosper in the long term through collective concerns and actions, not by solely selfish pursuits.[20]

Danger signals

The stories of the sardine ball and the penguin huddle suggest that it is dangerous to be on the social perimeter. Living on the perimeter threatens the lives and genetic legacy of humans as well. Epidemiological studies have found that social isolation is associated not only with lower levels of happiness and well-being, but also with broad-based morbidity and mortality.[21] Moreover, humans are such meaning-making creatures that perceived social isolation is at least as important a predictor of adverse outcomes on human health and well-being as objective social isolation.[22]

Writers may spend long periods by themselves, but the envisioned readers make this time feel anything but isolated. High school graduates who leave family and friends for the first time to attend college, on the other hand, typically experience intense feelings of social isolation even though they are physically around more people than before they left home. Avshalom Caspi and colleagues[23] found that *perceived* social isolation in adolescence and young adulthood predicted how many cardiovascular risk factors (such as body mass index, waist circumference, blood pressure, and cholesterol) were elevated in young adulthood, and that the number of developmental occasions (childhood, adolescence, and young adulthood) at which participants felt socially isolated predicted the number of elevated risk factors in young adulthood.

Perceived social isolation is known more colloquially as loneliness, which in early scientific investigations was depicted as "a chronic distress without redeeming features."[24] Loneliness may feel like a painfully miserable, hopeless, and worthless state, but we have found it has a specific structure and a valuable adaptive function.

Research on the ways in which people describe themselves when asked the question "Who are you?" reveals three basic dimensions[25]:

1. A *personal, or intimate, self,* the "you" of your individual characteristics

2. A *social or relational self,* which is who you are in relation to the people closest to you—your spouse, kids, friends, and neighbors

3. A *collective self,* the you that is the member of a certain ethnic group, has a certain national identity, belongs to certain professional or other associations, or is a member of the fan club for certain sports teams

Similar to the relational self, this part of the self is social, but what makes this self distinct is that these are broader social identities, linked to larger social groups rather than individual members of the groups. When we examined the dimensions underlying loneliness and social connectedness, we found the same three basic dimensions[26]:

1. *Intimate connection or isolation* refers to the perceived presence or absence of anyone in your life who affirms you as a valued person.

2. *Relational connection or isolation* refers to the perceived presence
or absence of quality friendships or family connections.

3. *Collective connection or isolation* refers to the perceived presence
or absence of a meaningful connection with a group or social stimulus (such as a school or team) beyond other individuals. When
you perceive that you are part of a valued group (collective connection), for instance, you may be more inclined to agree with
other group members, even on beliefs that may seem irrational,
than when you are thinking of yourself as a unique individual.

Given that human survival and prosperity depends on inclusion in
and participation with a social group—especially in evolutionary time,
when food was scarce and dangers were common—there is an adaptive
benefit to having the strong and aversive response of loneliness when an
individual feels his or her social connections might be weakening or
broken, just as there is a benefit to having aversive signals for other conditions critical for survival. Hunger, thirst, and pain have evolved as aversive signals to prompt organisms to change their behavior in a way that
protects individuals and promotes the likelihood that their genes will
make their way into the gene pool. The social pain of loneliness has
evolved similarly—to serve as a signal that one's connections to others
are weakening and to motivate the repair and maintenance of the connections to others that are needed for our health and well-being and for
the survival of our genes.[27] Physical pain is an aversive signal that evolved
to motivate one to take action that minimizes damage to one's physical
body. Loneliness is an aversive signal that evolved to motivate one to take
action that minimizes damage to one's social body.

People differ dispositionally in their sensitivity to the pain of social
disconnection (feelings of loneliness[28]) just as people differ in sensitivity
to physical pain. Ostracism or objective isolation in most species is associated with an early death.[29] In humans, the chronic feeling of social isolation, even when the person remains among the protective embrace of
others, is associated with significant mental and physical disorders.[30]
Chronic hunger, thirst, and pain can also have deleterious effects
because, like loneliness, their adaptive value lies in their effects as acute
signals, not as chronic conditions. The opposite of feeling hunger, thirst,
pain, or loneliness is feeling *normal,* and this is the state in which most
people exist most of the time.

The social connections formed by humans need not be based on genetic similarities. The human species has evolved the capacity for and motivation to form relationships not only with other individuals, but also with groups (such as a Chicago Cubs or a Boston Red Sox fan) and non-human entities (such as through anthropomorphism).[31] Team spirit and school spirit are familiar notions, and although team or school spirit refers to an invisible influence, it is an invisible influence that is no less open to rigorous scientific inquiry than are the invisible influences of gravity or magnetic flux. In the cases of team and school spirit, this influence represents a specific form of social connection between an individual and an emergent structure.

Perceived social connections are abstractions that can transcend time and space. People may feel a connection with their ancestors or family heritage even if they are the only remaining descendant, just as people can perceive personal connections with pets, cars, television characters, celebrities, and unseen spiritual entities with whom they do not actually interact. A potent component of spirituality (that does not depend on a specific religion) is the feeling of connection and purpose that comes from forming a relationship with a higher being. A simple by-product of our selfish genes and social brains may be our search for meaning in and connection with broader organizations or beings.

Conclusion

In 1939, the astrophysicist Sir Arthur Eddington published a book entitled *The Philosophy of Physical Science*.[32] In it he describes a hypothetical scientist who sought to determine the size of the various fish in the sea. The scientist began by weaving a 2-inch mesh net and setting sail across the seas, repeatedly sampling catches and carefully measuring, recording, and analyzing the results of each catch. After extensive sampling, the scientist concluded that there were no fish smaller than 2 inches in the sea. The moral of Eddington's analogy is twofold. First, scientific instruments are shaped by people's intuitive theories of the phenomena being investigated. Second, once developed, scientific expectations and instruments shape data and theory in ways more powerfully and fundamentally than are often appreciated.

What is the relevance of this to the story of the social nature of humankind? Our research findings have led me to believe that we all have made Eddington's error in the way we have thought about, studied, and tried to deal with an invisible force that motivates us to seek and maintain our connection with others—including the loneliness we feel when important social connections are threatened or absent. Historically, the scientific perspective on loneliness was not only that it was a painful and miserable state, but also that it was an aversive state with no redeeming features. All one needs to do is to reflect on the last time one felt terribly lonely, and one can appreciate the seemingly self-evident truth of this characterization. But as Eddington's story shows us, the obvious and intuitive can sometimes be very misleading. It is now widely recognized that many structures and processes of the mind operate outside of awareness, with only the end products sometimes reaching awareness.

Humans have evolved to seek connections with and validation from other minds, and these social connections represent an important set of invisible forces operating on our brain and biology. The need for social connection extends beyond kin relations and beyond face-to-face relations to include felt connections with superorganismal entities such as teams, political parties, nations, and God. The unseen forces compelling these connections can be quantified and investigated objectively independent of one's spiritual beliefs. Underlying these aspirations are selfish genes that have produced a social brain that activates reward regions of the brain when we cooperate effectively with others[33] or punish the perpetrators of social exploitations,[34] and that activates the pain matrix in the brain when we feel rejected by others.[35] When people feel socially isolated (lonely) compared to when they do not feel lonely, they are more likely to not only perceive nonhuman objects as humanlike, but also to believe in the existence of God.[31, 36] To understand the full capacity of forces operating on humans, we need to appreciate not only the memory and computational power of the brain, but also its capacity for representing, understanding, and connecting with other individuals and with the emergent structures, fictional and real, that the brain can represent. We need to recognize that we have evolved a powerful, meaning-making *social* brain and a need for social connection.

Endnotes

1. R. Dawkins, *The Selfish Gene* (Oxford: 1976).

2. G. C. Williams, "Natural Selection, the Costs of Reproduction, and a Refinement of Lack's Principle," *The American Naturalist* 100 (1966): 687–690.

3. D. S. Wilson and E. O. Wilson, "Evolution 'for the Good of the Group,'" *American Scientist* 96 (2008): 380–389.

4. B. Hölldobler and E. O. Wilson, *The Superorganism: The Beauty, Elegance, and Strangeness of Insect Societies* (New York: W. W. Norton, 2008).

5. S. Bowles, "Did Warfare Among Ancestral Hunter-Gatherers Affect the Evolution of Human Social Behaviors?" *Science* 324 (2009): 1293–1298.

6. J. Haidt, J. Patrick Seder, and S. Kesebir, "Hive Psychology, Happiness, and Public Policy," *The Journal of Legal Studies* 37 (2008): S133–S156.

7. D. S. Wilson, *Evolution for Everyone: How Darwin's Theory Can Change the Way We Think About Our Lives* (New York: Delacorte Press, 2007).

8. R. Dunbar, *The Human Story* (London: Faber and Faber, 2004).

9. R. I. M Dunbar and S. Shultz, "Evolution in the Social Brain," *Science* 317 (207): 1344–1347.

10. S. Beckerman, P. I. Erickson, J. Yost, J. Regalado, L. Jaramillo, C. Sparks, M. Iromenga, and K. Long, "Life Histories, Blood Revenge, and Reproductive Success Among the Waorani of Ecuador," *Proceedings of the National Academy of Sciences of the United States of America* 106 (2009): 8134–8139.

11. J. T. Cacioppo, D. G. Amaral, J. J. Blanchard, J. L. Cameron, C. Sue Carter, D. Crews, S. Fiske, T. Heatherton, M. K. Johnson, M. J. Kozak, R. W. Levenson, C. Lord, E. K. Miller, K. Ochsner, M. E. Raichle, M. Tracie Shea, S. E. Taylor, L. J. Young , and K. J. Quinn, "Social Neuroscience: Progress and Implications for Mental Health," *Perspectives on Psychological Science* 2 (2007): 99–123.

12. Z. R. Donaldson and L. J. Young, "Oxytocin, Vasopressin, and the Neurogenetics of Sociality," *Science* 322 (2008): 900–904.

13. C. O. Lovejoy, "Reexamining Human Origins in Light of Ardipithecus Ramidus," *Science* 326 (2009): 74e1–74e8.

14. H. Pakkenberg, "The Number of Nerve Cells in the Cerebral Cortex of Man," *The Journal of Comparative Neurology* 128 (1966): 17–20.

15. G. G. Berntson, G. J. Norman, and J. T. Cacioppo, "Evaluative Processes," in *Handbook of Neuroscience for the Behavioral Sciences* (New York: Wiley, 2009).

16. N. Emler, "Gossip, reputation and adaptation," in R. F. Goodman and A. Ben-Ze-ev (eds.), *Good Gossip* (Lawrence, Kans.: University of Kansas Press, 1994), p. 34–46.

17. D. Kahneman, A. B. Krueger, D. A. Schkade, N. Schwarz, and A. A. Stone, "A Survey Method for Characterizing Daily Life Experience: The Day Reconstruction Method," *Science* 306 (2004): 1776–1780.

18. E. Berscheid and H.T. Reis, "Attraction and Close Relationships," in D. Gilbert, S. T. Fiske, and G. Lindzey (eds.), *The Handbook of Social Psychology* (New York: McGraw-Hill, 1998): 193–281.

19. M. Gansberg, "Thirty-Eight Who Saw Murder Didn't Call the Police," *New York Times* (1964).

20. O. Gurerk, B. Irlenbusch, and B. Rockenbach, "The Competitive Advantage of Sanctioning Institutions," *Science* 312 (2006): 108–111.

21. J. S. House , K. R. Landis, and D. Umberson, "Social Relationships and Health," *Science* 241 (1988): 540–545.

22. J. T. Cacioppo and B. Patrick, *Loneliness: Human Nature and the Need for Social Connection* (New York: W. W. Norton & Company, 2008).

23. A. Caspi, H. Harrington, T. E. Moffitt, B. J. Milne, and R. Poulton, "Socially Isolated Children 20 Years Later: Risk of Cardiovascular Disease," *Archives of Pediatrics & Adolescent Medicine* 160 (2006): 805–811.

24. R. S. Weiss, *Loneliness: The Experience of Emotional and Social Isolation* (Cambridge, Mass.: MIT Press, 1973).

25. M. B. Brewer and W. Gardner, "Who Is This 'We'? Levels of Collective Identity and Self-Representations," *Journal of Personality and Social Psychology* 71 (1996): 83–93.

26. L. C. Hawkley, M. W. Browne, and J. T. Cacioppo, "How Can I Connect with Thee? Let Me Count the Ways," *Psychological Science* 16 (2005): 798–804.

27. J. T. Cacioppo, L. C. Hawkley, J. M. Ernst, M. Burleson, G. G. Berntson, B. Nouriani, and D. Spiegel, "Loneliness Within a Nomological Net: An Evolutionary Perspective," *Journal of Research in Personality* 40 (2006): 1054–1085.

28. D. I. Boomsma, G. Willemsen, C. V. Dolan, L. C. Hawkley, and J. T. Cacioppo, "Genetic and Environmental Contributions to Loneliness in Adults: The Netherlands Twin Register Study," *Behavior Genetics* 35 (2005): 745–752.

29. K. D. Williams, *Ostracism: The Power of Silence* (New York: Guilford Press, 2001).

30. J. T. Cacioppo and B. Patrick, *Loneliness: Human Nature and the Need for Social Connection* (New York: W. W. Norton & Company, 2008).

31. N. Epley, A. Waytz, S. Akalis, and J. T. Cacioppo, "When We Need a Human: Motivational Determinants of Anthropomorphism," *Social Cognition* 26 (2008): 143–155.

32. A. S. Eddington, *The Philosophy of Physical Science* (Cambridge, Mass.: Cambridge University Press, 1939).

33. J. Rilling, D. Gutman, T. Zeh, G. Pagnoni, G. Berns, and C. A. Kilts, "Neural Basis for Social Cooperation," *Neuron* 35 (2002): 395–504.

34. D. J. F. De Quervain, U. Fischbacher, V. Treyer, M. Schellhammer, U. Schnyder, A. Buck, and E. Fehr, "The Neural Basis of Altruistic Punishment," *Science* 305 (2004): 1254–1258.

35. N. I. Eisenberger, M. D. Lieberman, and K. D. Williams, "Does Rejection Hurt? An fMRI Study of Social Exclusion," *Science* 302 (2003): 290–292.

36. N. Epley, A. Waytz, and J. T. Cacioppo, "On Seeing Human: A Three-Factor Theory of Anthropomorphism," *Psychological Review* 114 (2007): 864–886.

From inclusive fitness
to spiritual striving

The notion of "selfish genes" (and, by extension, selfish organisms) was popularized in Richard Dawkins's 1976 book by that title. Not long afterward, an article appeared in *Science* that presented evidence that the most vicious members of a warlike tribe in South America had the most wives and children. The underlying notion was one of (genetic) survival of the fittest: Those warriors who were particularly vicious were more likely to contribute their genes to the gene pool. Methodological objections have left this an open question, however, and new evidence now exists that calls this interpretation into question: The most aggressive warriors may have more children, but they have *lower* indices of reproductive success than their milder brethren, perhaps partly because the most aggressive warriors and their offspring are also more likely to be the targets of revenge killings. This new data is entirely consistent with Cacioppo's argument that the content of the human gene pool has more to do with the reproductive success of one's offspring than one's own reproductive success. Cacioppo argued further that this genetic selection resulted in a social brain that seeks meaning and connection with individuals and with social entities (such as groups) that extend beyond other individuals.

In the next chapter, "Science, religion, and a revived religious humanism," theologian Don Browning also embraces the concept of inclusive fitness and, through the writings of Thomas Aquinas, shows how religion serves the human need for meaning and connection through the ethics they advocate, the congregations they form, the institutions they represent, and the God they serve. In his view, religion serves to extend love and connection beyond kin. Browning further argues that new developments in the sciences and long-standing traditions in theology constitute fertile ground on which to build new and testable hypotheses regarding our fundamental human nature.

3 *

Science, religion, and a
revived religious humanism

For more than 150 years, there has been a vital, and often contentious, dialogue between science and religion. In recent years, new energy and fresh public interest have been injected into this conversation. This was largely spurred by the new insights into religion and ethics achieved by collaboration between evolutionary psychology and cognitive and social neuroscience.

* The lead author is Don Browning, Ph.D., the Alexander Campbell Professor of Religious Ethics and the Social Sciences, Emeritus, Divinity School, University of Chicago. He has interests in the relation of the social sciences to religious ethics for the purpose of addressing various challenges facing modern life. His books include *Generative Man* (1973, 1975; National Book Award Finalist, 1974), *Religious Thought and the Modern Psychologies* (1987, 2004), the coauthored *From Culture Wars to Common Ground: Religion and the American Family Debate* (1997, 2000), *Christian Ethics and the Moral Psychologies*

What are the likely social consequences of this new interest in the relation of science and religion? There are at least three possible answers. One might be the new atheism exemplified by the writings of Richard Dawkins, Daniel Dennett, Sam Harris, and Christopher Hitchens.[1] In this approach, the alleged defective thinking of the world religions is exposed, and a worldview and way of life based strictly on science are offered as replacement. A second option might be the return of a hegemonic dominance of religion over science. However, polarizing rhetoric from advocates for the exclusive interpretive priority of either science or religion has long since ceased to be culturally or academically productive. Instead, through dialogue about common issues, scientific and theological thinkers may pose questions that lead to more sophisticated inquiry in both fields. Confidence in the productive possibilities of reciprocal questioning is a hallmark of the long tradition known as religious humanism. In this chapter, I illustrate the potential contribution of religious humanism by bringing recent psychological research into dialogue with the religious concept of love.

What would this religious humanism be like? The major world religions would remain visible and viable as religious movements. But the contributions of science would help these religions refine their interests in improving the health, education, wealth, and overall well-being of their adherents. In addition, the sciences would help them refine their grasp of the empirical world about which they are, like humans in general, constantly making judgments, predictions, and characterizations. In my vision, the attitude of scientists toward religion would first be phenomenological; they would first attempt to describe and understand religious beliefs, ethics, and rituals in their full historical context. But their interest in explaining some of the conditions that give rise to religious

(2006), *and Equality and the Family* (2007). He coedited *Sex, Marriage, and Family in the World Religions* (2006), *American Religions and the Family* (2006), *Children and Childhood in American Religions* (2009), and *Children and Childhood in World Religions* (2009). He is the coprincipal investigator with Jean Bethke Elshtain of a grant on the New Science of Virtue project.

In this essay, Browning acknowledges the antagonistic relationship that can be found between science and religion, but he proposes that the dialogue between science and religion can now be conducted on philosophical grounds that promote a new religious humanism that will honor the core ideas of the great religions; refine their view of nature; and increase the values of health, wealth, education, and general well-being.

phenomena would not be inhibited by either religion or the wider society. There are several different approaches to phenomenology. The perspective that I recommend follows the "critical hermeneutic phenomenology" of Paul Ricoeur. Ricoeur advocates beginning the study of religion with a phenomenology—a careful description—of the person's or community's words, symbols, metaphors, and narratives used to communicate the meaning of a religious experience or practice. This view assumes that we cannot describe experience directly because experience is always mediated by symbols and metaphors. But Ricoeur's phenomenology does not stop with a description of these meanings. It builds in a secondary place for science and explanation. It seeks through science to give explanatory accounts of the affects and motivations that humans bring to these words, symbols, and metaphors. If scientists followed Ricoeur's model, they would understand the importance of beginning with description, be hesitant to skip lightly over initial phenomenological meanings, appreciate yet grasp the limits of explanation, and be reluctant to plunge into speculations, such as those of the new scientific atheism, about the ultimate truth or falsity of religious phenomena.[2]

On the other hand, the religions themselves can contribute to the sciences. They can do this by offering hypotheses about how social and religious ideas, behaviors, and rituals can shape experience, even neural processes, often for the good. The religions can offer a more generous epistemology and ontology than science is inclined to find useful for the tight explanatory interests of the laboratory or scientific survey. This, too, might generate new hypotheses for scientific investigation. These would be some of the ground rules for how a dialogue between science and religion might stimulate a revived religious humanism.

Religious humanisms of the past

To speak of a revival suggests that there have been many expressions of religious humanism in the past when some form of science, philosophy, and religion creatively interacted. I limit myself to speaking primarily about Judaism, Christianity, and Islam. A synthesis between Greek philosophical psychology and Christianity can be found in the use of Stoic theories of desire by the apostle Paul,[3] the presence of Aristotle's family ethic—with its implicit psychobiology—in the household codes of Ephesians and Colossians,[4] and the Gospel of John's identification of Jesus with the Platonic and Stoic idea of the preexistent "Word."[5] A more intentional religious humanism can be found in Augustine's use of the neoplatonic Plotinus, especially in the philosophical psychology of remembrance developed in his *Confessions*.[6] But the most dramatic example of a religious humanism that spread simultaneously into Judaism, Christianity, and Islam can be found when the lost texts of Aristotle were discovered, translated, and appropriated by scholars from these three religions who worked at the same tables in Islamic libraries in Spain and Sicily during the ninth and tenth centuries. In his timely book *Aristotle's Children,* Richard Rubinstein tells the story well.[7] This gave rise to forms of Aristotelian religious humanism in the works of Thomas Aquinas in Christianity, Maimonides in Judaism, and Averroës in Islam. On the American scene, one sees another form of Christian humanism in the synthesis of philosophical pragmatism, with all its influence from Darwin, and expressions of liberal Christianity and the social gospel movement.[8]

Religious humanisms have not always flourished and are subject to attacks from both fundamentalists and scientific secularists. They need constant updating and vigorous intellectual development. But at their best, they make it possible for societies to maintain strong religious communities and integrate symbolic umbrellas that protect the productive interaction of the scientific disciplines with the wider cultural and religious life.

An example: the *agape, caritas,* and *eros* debate

Few words in the English language have such a range of everyday meanings and of serious philosophical and theological consideration as the word *love.* For this reason, it is an excellent candidate for scientific investigation that has potential benefits for religious practice and everyday life. Although some theologians have sought to create a sharp division between "Christian love" and all other forms of love, the tradition of religious humanism proposes that science clarifies the workings of love in human societies and that religion extends the scope of love beyond its most immediate domain of kinship.

Three major tensions exist in theological discussions of Christian love. They center around the two Greek words *agape* and *eros* and the

Latin word *caritas*. A famous book titled *Agape and Eros,* written by the Swedish theologian Anders Nygren, traced the debate through Christian history.[9] Nygren believed that the truly normative and authentic understanding of Christian love is found in the word *agape,* the Greek word used for Christian love in the New Testament. It refers to a kind of self-giving, even self-sacrificial, love that is possible only by the grace of God.[10] Nygren was particularly interested in arguing that Christian love did not build on what the Greek philosophers called *eros.* He claimed *eros* refers to the natural desires of humans to have and unite with the goods of life. This includes the goods of health, wealth, affiliation, and pleasure, but it also includes the higher goods of beauty and truth. Nygren's point, however, was that *Christian love does not build on or incorporate eros*—the natural aspirational strivings of humans. He believed he found this view of Christian love in the New Testament (especially the writing of the apostle Paul) and Martin Luther, the giant of the Protestant Reformation.

Nygren was particularly interested in dismantling the classical medieval Roman Catholic view of Christian love that was often summarized with the English word *charity* or the Latin word *caritas.* Why did Nygren oppose the *caritas* view of Christian love? The answer is that the meaning of love as *caritas* did exactly what Nygren thought Paul and Luther, his theological heroes, did not do. In the classic Roman Catholic view, love as *caritas* builds on *eros. Caritas* was seen to include natural desires for health and affiliation. But the *caritas* view of love also held that religious education and God's grace built on and expanded these natural inclinations to entail a self-giving benevolence to others, even strangers and enemies—an idea so central to the concept of Christian love.

All this seemed too naturalistic for Nygren. It seemed to downplay the importance of God's transforming grace. He joined other European neoorthodox theologians of his day, such as Karl Barth and Rudolph Bultmann, in cutting off Christian love from *eros,*[11] which, in effect, was to cut off Christian love from nature and desire—the very things scientists tend to study. Beginning with Nygren's strong view of *agape* and the strong supernaturalism of both Nygren and Barth, there was little room in these mid-twentieth-century Protestant trends for a productive dialogue between Christian ethics and the new scientific advances in moral psychology, evolutionary psychology, and neuroscience.

At the same time, however, breakthroughs in these very disciplines have led to a new reassessment of the Catholic *caritas* model of Christian love. But before I review in more detail how this model worked, especially in the thought of the great medieval Roman Catholic theologian Thomas Aquinas, let me review some of the moral implications of insights into kin altruism and inclusive fitness emerging today from evolutionary psychology and social neuroscience.

Moral implications of kin altruism and inclusive fitness

As is well known, the idea of inclusive fitness was first put forth in 1964 by William Hamilton.[12] Hamilton's view of inclusive fitness holds that living beings struggle not only for their individual survival, but also for the survival of offspring and kin who carry their genes. Their altruism is likely to be proportional to the percentage of their genes that others carry. This insight was further developed by the concept of parental investment. Ronald Fisher and Robert Trivers defined it as "any investment by the parent in an individual offspring that increases the offspring's chance of surviving ... at the cost of the parent's ability to invest in other offspring."[13] These insights were at the core of the emerging field of sociobiology and were first brought to the wider public attention by E. O. Wilson's *Sociobiology: The New Synthesis.*[14]

But the moral implications of the concept of inclusive fitness, parental investment, and kin altruism have received competing interpretations. In his *The Selfish Gene,* Richard Dawkins turned these ideas into a defense of philosophical ethical egoism and argued that all altruistic acts are disguised maneuvers to perpetuate our own genes.[15] But there are other interpretations. Social neuroscientist John Cacioppo interprets our motives toward inclusive fitness and kin altruism as the core of human intergenerational care and the vital link between sociality and spirituality. In cooperation with his colleagues, his research on loneliness uses evolutionary theory on inclusive fitness to order many of his findings. From the perspective of this model of basic human motivations, loneliness can be seen as a condition that "promotes inclusive fitness by signaling ruptures in social connections and motivates the repair or replacement of these connections."[16] According to his interpretation of inclusive fitness, our gene continuity is not assured by simply having our own children. Our children also must have children. And this is a challenge entailing long-term expenditures of energy. To account for this, Cacioppo writes something about human infant dependency that is very close to what both the Greek philosopher Aristotle and the medieval Roman Catholic theologian Thomas Aquinas set down many centuries earlier. Cacioppo says,

> For many species, the offspring need little or no parenting to survive and reproduce. *Homo sapiens,* however, are born to the longest period of abject dependency of any species. Simple reproduction, therefore, is not sufficient to ensure that one's genes make it into the gene pool. For an individual's genes to make it to the gene pool, one's offspring must survive to reproduce. Moreover, social connections and the behaviors they engender (e.g., cooperation, altruism, alliances) enhance the survival and reproduction of those involved, increasing inclusive fitness.[17]

According to this view, the twofold interaction between inclusive fitness and the long period of infant dependency has shaped humans over the long course of evolution into the social and caring creatures we are. Sociality is a fundamental characteristic of humans, and, according to Cacioppo, spirituality, in its various forms, is an extension of sociality. Religion is generally, although not always, good for our mental and physical health—our heart, blood pressure, self-esteem, and self-control— just as having good friends and family or not being lonely are also good for our well-being.[18] Cacioppo and colleagues do not equate sociality and religion; they are fully aware that religions are complex phenomena with many different doctrinal, ethical, ritual, organizational, personal, and social features that require either rigorous experimental or clinical population studies to sort out, even from the perspective of how they affect health. Nonetheless, he seems to hold that the sociality that most religions offer is a key reason for their efficacy in human well-being.

Does Christian love build on health?

But my concern is the topic of Christian love, not simply Christianity's contribution to mental and physical health. Although Jesus is said to have

performed miracles of health, offering health in this world has never
been at the core of Christianity or, for that matter, the other Abrahamic
religions of Judaism and Islam. Bringing to maturity loving and self-
giving persons has been the primary concern of Christianity, whether or
not this contributes to health and well-being. But the question is, as I
elaborated earlier, does Christian love build on *eros*—our strivings for
health and other goods—or come exclusively from some transnatural
source, as Nygren believes the normative tradition taught? And did
Christianity ever identify *eros* and our deepest motivations with something like inclusive fitness and kin altruism?

Let me start with Aquinas. In the "Supplement" to his *Summa Theologica III*, Aquinas follows Aristotle and the Roman natural-law theorist
Ulpian in asserting that humans share with all animals an inclination to
have offspring.[19] Having said this, he then introduces a very modern-
sounding commentary on the uniqueness to humans of the long period
of infant dependency. Notice the similarity of his argument to the words
of Cacioppo. Aquinas writes

> Yet nature does not incline thereto in the same way in all ani-
> mals; since there are animals whose offspring are able to seek
> food immediately after birth, or are sufficiently fed by their
> mother; and in these there is no tie between male and female;
> whereas in those whose offspring needing the support of both
> parents, although for a short time, there is a certain tie, as may
> be seen in certain birds. In man, however, since the child
> needs the parents' care for a long time, there is a very great tie
> between male and female, to which tie even the generic nature
> inclines.[20]

Although this quote contains a description of how family formation
emerges at the human level, there is an implicit argument for both the
fact of human infant dependency and what we today call inclusive fitness. But these ideas are even more evident in the next quote, although
stated very much from the male point of view, a habit typical of his time.
Aquinas says, "Since the natural life which cannot be preserved in the
person of an undying father is preserved, by a kind of succession, in
the person of the son, it is naturally befitting that the son succeed in the
things belonging to the father."[21] Aquinas's main source for this insight
was Aristotle's *Politics*. In one place, Aristotle wrote, "In common with

other animals and with plants, mankind have a natural desire to leave behind them an image of themselves."[22]

However, in both Aristotle and Aquinas, such claims were not just about the importance of kin continuity; they were statements about the origin and need of long-term investments by parents at the human level. In contrast to his teacher Plato, who, in his *Republic,* had advocated removing children from their biological parents in an effort to overcome the civil frictions created by nepotism,[23] Aristotle counters with an assertion about the origins of human care. Aristotle wrote, "That which is common to the greatest number has the least care bestowed upon it." He believed that, in Plato's state, "love will be watery.... Of the two qualities which chiefly inspire regard and affection—that a thing is your own and that it is your only one—neither can exist in such a state as this."[24] This is an assertion about the importance of kin altruism in human care.

Although Aquinas saw these natural inclinations as important for the formation of long-term human care, he believed that they were not sufficient for mature parental love. Powerful social, cultural, and, indeed, religious reinforcements were also necessary for stable parental investment to be realized. Once again, this is, according to Aquinas, because of the many long years of human childhood dependency; human children need their parents for a very long period of time and over the course of many contingencies and challenges. This led Aquinas to develop his theology of marriage as a way of consolidating and stabilizing parental commitment, especially paternal commitment, to their dependent children.[25]

Although neither Aristotle nor Aquinas presented the full intergenerational scope of Cacioppo's interpretation of kin altruism and inclusive fitness—that it must extend to our children's children and not just our own—both perspectives comprehended the interlocking nature of kin altruism and the wellspring of care, long-term human commitment, and, hence, some of the rudimentary energies of human morality.

Of course, Aquinas and those who followed him supplemented these naturalistic observations with additional epistemological presuppositions that may seem strange to scientists. These included the idea that God works through nature and grace, hence God is present in the kin altruistic inclinations of parents and grandparents. He also assumed that, for kin altruism to be stable, the additional social reinforcements of institutional marriage and God's strengthening grace and forgiveness were also needed. In addition, he held—and Christianity has always taught—that

Christian love includes more than kin altruism and the care of our familial offspring; it must include the love of neighbor, stranger, and enemy,
even to the point of self-sacrifice. For the Christian, this was made possible by the idea that God was the creator of all humans, and hence each
person was a child of God and made in God's image *(imago dei)*. For this
reason, as Kant would say on different grounds, each individual should
be treated "always as an end and never as a means only."[26] In Aquinas's
view, acting on this belief, and with the empowering grace of God, made
it possible for Christians *to build on, yet analogically generalize, their kin
altruism to all children of God, even those beyond the immediate family,
their own children, and their own kin.* These wider assumptions may be
beyond the competence of science to assess. They entail a step toward
metaphysical speculation of the kind science would do better to avoid.
Nonetheless, in the view of Christian love developed in Aquinas, the
seeds of a religious humanism—in this case, a Christian humanism—
began to form.

I have tried to illustrate how insights from Aristotle and Aquinas can
join with insights from evolutionary psychology and social neuroscience
to refine the Christian understanding of love. In pursuing this course, I
join the work of Stephen Pope and others in presenting this option.[27] The
kind of Christian humanism found in Aquinas makes it possible for Christianity to be enriched by the modern sciences of human nature. Aquinas's
view is strikingly different from Nygren's representation of Paul and
Luther when Nygren contends that Christian *love does not build on our
own natural energies, but "has come to us from heaven."* [28] Or again, it is
very different from Nygren's view when he writes that the Christian is
"merely the tube, the channel through which God's love flows."[29] The
complete discontinuity in this statement between the downward love of
God and the natural extension to nonkin of human kin-altruistic impulses
is stunning. Such a view as Nygren's precludes the possibility of a religious humanism of the kind I have been describing. And it eliminates the
possibility of the refinements to religious views of human nature that the
conversation between religion and science can offer.

Conclusion

My argument has been that a revived religious humanism can come
about through the dialogue between religion and science, particularly

between religion and the psychological sciences. I have illustrated this with the issue of love in Christianity. I believe my argument could be illustrated with other religions, especially the Abrahamic religions of Judaism and Islam. As Aristotle's influence created a kind of religious humanism in these religions in the past, the broader dialogue between science and religion may be able to do this for them in the future.

But the contributions will not simply flow from science to religion. Even in this short chapter, a question for science to investigate has arisen: How do religious and metaphysical beliefs extend the impulse of natural kin altruism, if at all? This goes beyond the issue of the relation of religion to health. It raises the question of the relation of religion to expansive love for the distant other. This is a good question that comes from taking the claims of religion seriously and is an example of how religion can continue to feed and challenge scientific inquiry.

Endnotes

1. Richard Dawkins, *The God Delusion* (Boston, Mass.: Houghton Mifflin Co., 2006); Daniel Dennett, *Breaking the God Spell* (New York: Viking, 2006); Sam Harris, *Letter to a Christian Nation* (New York; Alfred Knopf, 2006); and Christopher Hitchens, *God Is Not Great* (New York: Twelve, 2007).

2. For a discussion of these distinctions between different forms of phenomenology, see Paul Ricoeur, *Hermeneutics and the Human Sciences* (Cambridge, Mass.: Cambridge University Press, 1981), p. 63–100.

3. Will Deming, *Paul on Marriage and Celibacy* (Cambridge, Mass.: Cambridge University Press, 1995); and Troels Engbert–Perdersen, *Paul and the Stoics* (Louisville, Ky.: Westminster John Knox, 2000).

4. Aristotle, *Nicomachean Ethics* (New York: Random House, 1941), Bk. VIII, ch. 10.

5. G.A. Buttrick, *The Interpreter's Bible: Luke and John, Vol. 8* (Nashville, Tenn.: Abingdon Press, 1952), p. 465.

6. Peter Brown, *Augustine of Hippo* (Berkeley, Calif.: University of California, 1969), p. 178.

7. Richard Rubinstein, *Aristotle's Children* (New York: Harcourt, 2003).

8. Edward Scribner Ames, *Religion* (Chicago: Holt, 1929).

9. Anders Nygren, *Agape and Eros* (Philadelphia.: Westminster Press, 1953).

10. *Ibid.*, p. 57, 121–122.

11. *Ibid.*, p. 101.

12. William Hamilton, "The Genetical Evolution of Social Behavior II," *Journal of Theoretical Biology* 7 (1964): 17–52.

13. Ronald Fisher and Robert Trivers, "Parental Investment and Sexual Selection," B. Campbell, ed., *Sexual Selection and the Descent of Man* (Chicago: Aldine Publishing Co., 1972), p. 139.

14. E. O. Wilson, *Sociobiology: The New Synthesis* (Cambridge, Mass.: Harvard University Press, 1975).

15. Richard Dawkins, *The Selfish Gene* (Oxford: Oxford University Press, 1976).

16. See Chapter 2 of this book, "The social nature of humankind."

17. *Ibid.*, p. 5.

18. J.T. Cacioppo and M.E. Brandon, "Religious Involvement and Health: Complex Determinism," *Psychological Inquiry* 13, no. 3 (2002): 204–206.

19. Larry Arnhart, "Thomistic Natural Law as Darwinian Natural Right," Ellen Frenkel, Fred Miller, and Jeffrey Paul eds, *Natural Law and Modern Moral Philosophy* (Cambridge, Mass.: Cambridge University Press, 2001), p. 1–2.

20. Thomas Aquinas, *Summa Theologica III*, "Supplement," q., 41.1.

21. Thomas Aquinas, *Summa Contra Gentiles* (London: Burns, Oates, and Washburn, 1928), 3, ii, p. 115.

22. Aristotle, "Politics," Richard McKeon ed., *The Basic Works of Aristotle* (New York: Random House, 1941), I, i.

23. Plato, *The Republic* (New York: Basic Books, 1968), Bk. 5, p. 461–465.

24. Aristotle, "Politics," Bk. 2, chpt. 3.

25. Aquinas, *Summa Contra Gentiles*, 3, ii, p. 115

26. Immanuel Kant, *Foundations of the Metaphysics of Morals* (New York: Bobbs–Merrill Co., 1959), p. 49.

27. Stephen Pope, "The Order of Love in Recent Roman Catholic Ethics," *Theological Studies* 52 (1991): 255–288.

28. Anders Nygren, *Agape and Eros* (Philadelphia: Westminster Press, 1953), p. 734.

29. *Ibid.*, p. 735.

The status of the body politic and the status of the body itself

In the long history of conversations between science and religion, start-ing points matter. As Don Browning demonstrates through the example of Thomas Aquinas, when religions start with a desire to understand human behavior, such as the long-term human commitment between parent and child, they recognize that science might illuminate and refine that understanding. And religion, in turn, might shape social institutions that not only enhance the human drive toward social connection, but also imaginatively extend its influence beyond direct kinship to influence ethical relations with neighbor and stranger. By starting with a shared interest in understanding what Browning calls "the rudimentary energies of human morality," creative conversation between science and religion thus prompts a religious humanism, in which religion partners with sci-ence to shape models of fulfillment for human sociality.

Like Browning, Louise Hawkley starts with the human quest for social connection. As humans mature, Hawkley observes, they proceed from childhood dependence not toward independence, but toward inter-dependence. But whereas Browning pursued the implications of interde-pendence for the body politic, Hawkley wants to know the consequences of interdependence for the physical body. Her research focuses on the interplay between psychological and physical factors in the human sense of social connectedness, and Hawkley finds that "feeling wanted and accepted and like one belongs are as vital to our existence as the air we breathe." A robust sense of social connection reverberates throughout the human body, and its absence—in loneliness—is likely to have long-term adverse effects on personal health.

4 *

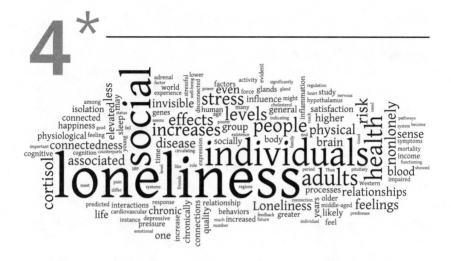

Health by connection: from social
brains to resilient bodies

We enter and leave the world alone, but at no time do we exist disconnected from others. The connection with the mother that begins *in utero* does not end with the physical severing of the umbilical cord, but continues in a lengthy dependence on the mother or primary caregiver(s) for food and safety. Years are needed for the infant to reach physical and

---

* The lead author is Louise C. Hawkley, Ph.D., a Senior Research Scientist (Assistant Professor), member of the Center for Cognitive and Social Neuroscience, and Associate Director of the Social Neuroscience Laboratory at the University of Chicago. Her research is concerned with the interplay of psychological and physiological factors, and includes the study of autonomic, neuroendocrine, immune, genetic, and behavioral processes that contribute to physical and mental health and well-being in individuals differing in perceived social connectedness. She has published numerous articles and chapters on perceived social

reproductive maturation—but as Cacioppo notes, even more important, these years are needed for the infant to develop the social and emotional skills necessary for survival in a complex social world. We graduate from infantile dependence not to independence, but to interdependence (cooperation, trust, reciprocity, and so on). That we are born to and for connection explains why feeling disconnected, isolated, and like we don't belong can be so painful. We call these feelings loneliness. Feelings of loneliness function like physical pain or hunger or thirst; they motivate us to alleviate the social pain and to repair our sense of connectedness. This is an important adaptive function of loneliness because people who feel connected fare much better than those who feel disconnected. They are not only happier, but also healthier than their more lonely counterparts. As we will see, the power of felt connectedness reverberates throughout the human body. As was aptly stated by Frederick Buechner, "You can kiss your family and friends good-bye and put miles between you, but at the same time you carry them with you in your heart, your mind, your stomach, because you do not just live in a world but a world lives in you."[1] We tend to take for granted our sense of social connectedness, but that should not blind us to its powerful, albeit invisible, influence on our lives. Its impact is best exposed when we observe the effects of its absence (loneliness) on physical and mental health and well-being.

Loneliness is a health risk, but how?

For research purposes, loneliness is typically measured on a continuum that ranges from not at all lonely (socially connected) to very lonely. It is

isolation (loneliness) and its antecedents and consequences in young and aging adults.

Hawkley finds awe in the significance of what it means to be a social species. Atul Gawande wrote, "We are social not just in the trivial sense that we like company, and not just in the obvious sense that we each depend on others. We are social in a more elemental way: simply to exist as a normal human being requires interaction with other people" (*The New Yorker*, March 2009). Hawkley believes that we are social beings to our cellular core, and even that does not capture the full extent of our sociality. For these reasons, she is increasingly uncomfortable with the term *loneliness*, a term that is laden with popular definitions and understandings that only hint at the deeper significance of our social nature. As she shows, the study of loneliness actually reveals that feeling wanted and accepted and like one belongs are as vital to our existence as the air we breathe. Nothing kills like being denied a socially meaningful existence.

informative, however, to get a sense of the prevalence of loneliness when assessed as present or absent. Loneliness is a common experience; as many as 80% of people under 18 years of age and 30% of people over 65 years of age report being lonely at least sometimes. For most people, feelings of loneliness are situational and transient (such as during geographic relocation). For as many as 15–30% of the general population, however, loneliness is a chronic state, and it is among these individuals that loneliness wreaks its greatest havoc. In a study of children followed through young adulthood, those who were highly lonely at each of three measurement occasions (during childhood, during adolescence, and at 26 years of age) exhibited a significantly greater number of standard health risks. The chronically lonely individuals were more likely to have higher body mass index (BMI), elevated blood pressure, higher levels of total cholesterol, lower levels of "good" HDL cholesterol, greater concentrations of glycosylated hemoglobin (an index of impaired glucose metabolism), and poorer respiratory fitness than those who were lonely at only two or one of the measurement occasions.[2] In a study of older adults, loneliness predicted mortality during a three-year period, and increased mortality was explained by the fact that lonely individuals had more chronic diseases and functional limitations.[3] Higher rates of mortality in lonely individuals do not appear to be attributable to inadequate healthcare utilization: Even after accounting for the presence and severity of chronic illness, lonely individuals are actually more likely than nonlonely individuals to make use of health facilities and physicians.[4]

Most chronic diseases (such as hypertension, coronary artery disease, and diabetes) are the result of the interactive influences of genetic, environmental, and behavioral factors on physiological functioning. How do feelings of loneliness penetrate to a level that affects disease risk? Plausible pathways include poor health behaviors, stress-related processes, restorative "antistress" processes, and even differences in patterns of gene activity. In general, physiological systems exhibit redundancies and compensatory processes that minimize the immediate health effects of adverse heritable, environmental, and behavioral factors. However, subtle changes in these predisease pathways can be detected prior to the onset of manifest disease and may indicate the beginnings of a steeper downward trajectory in resilience.[5]

Take health behaviors, for instance. Major risk factors for disease in Western society include high-calorie, high-fat diets and sedentary lifestyles, each of which contributes to being overweight or obese. Feelings of loneliness have been associated with greater incidence of these pre-dominantly lifestyle risk factors. In a large cross-sectional survey of adults 18 years and older, the lonely group had a higher mean BMI and a greater proportion of overweight or obese individuals than the nonlonely group did. Loneliness has been associated with lower levels of physical activity in every age group from grade school to middle-age adults. In the latter study, lonely individuals were also more likely to become inactive over time. Changes in health status also predicted an increased likelihood of becoming inactive, but the effects of loneliness were independent of changes in health status. Similarly, individuals with fewer social connec-tions (a smaller social network) were less likely to be physically active, but the effects of loneliness on physical activity did not depend on the size of the social network. The invisible force of loneliness seems to play a unique role in this particular predisease pathway.

Another commonly cited risk factor for disease is stress. In reality, "stress" refers to a family of predisease pathways, each of which may be vulnerable to the influence of lonely feelings. Loneliness is itself a source of stress, but lonely individuals also differ in their exposure to stressful events and circumstances. This is less evident in young adults than it is in older adults in whom loneliness was associated with having experienced a greater number of stressful life events in the past year (such as a death in the family, marital crisis, or change in employment status) and more sources of chronic stress (such as employment stress or marital stress).[6] In addition, lonely individuals perceive life as more stressful and less gratifying than their socially connected counterparts, even when objec-tive indications are that lonely and nonlonely individuals do not differ in the types of activities and behaviors they engage in on a daily basis. Good-quality social interactions typically ameliorate feelings of stress, but because lonely people perceive their interactions to be less positive than those of nonlonely people, they fail to derive the same benefit. Good coping strategies can also ameliorate feelings of stress, but lonely individuals are more likely to respond to stress with pessimism and avoidance than with optimism and active engagement. And to add insult to injury, loneliness increases sensitivity to and surveillance for social threats. Anxiety, low self-esteem, and fear of negative evaluation elicit self-defensive behaviors and effectively tax cognitive resources that would normally be available to meet the demands of daily-life stress. Thus, what might naïvely be thought of as a circumscribed problem—the feeling of loneliness and isolation—may be seen by the sufferer as a world of inescapable threat.[7]

How might these cognitions and perceptions influence physiology and health? As Gary Berntson shows in Chapter 5, "Psychosomatic relations: from superstition to mortality," the brain regions involved in emotional and perceptual processes are intimately related to the brain regions involved in the regulation of physiological systems. This is particularly evident in alterations of the functioning of the cardiovascular system in lonely individuals. In young adults, this alteration is apparent in increased resistance to blood flow in small arteries throughout the body. Increased vascular resistance is a precursor and predominant contributor to age-related increases in systolic blood pressure (SBP), a major risk factor for cardiovascular disease. In middle-aged adults, SBP is significantly higher in lonely adults than in their nonlonely counterparts. Moreover, loneliness accelerates the rate of increases in SBP, indicating a faster decline in physiological resilience and a heightened risk for chronic cardiovascular disease.[8] It's as though loneliness accelerates the aging process.

By virtue of extensive interconnections among the brain, peripheral nervous systems, and endocrine glands, the feelings of isolation and loneliness have a broad and deep reach. The hypothalamus plays a key role in enabling communication from the brain to the periphery. Located in the lower central region of the brain, the hypothalamus receives neural input on brain and body states (such as pain, sadness, fear, and hunger) and, in response, signals brain regions that control the autonomic nervous system and the pituitary and adrenal glands. Signals to the autonomic nervous system permit modulation of heart rate, blood pressure, and numerous other factors that serve to maintain homeostasis. Signals to the pituitary gland (located at the base of the brain, just below the hypothalamus) prompt the release of hormones that ultimately permit modulation of almost every endocrine gland in the body, including the adrenal glands (one is situated on top of each kidney). The adrenal glands serve many functions, including producing and secreting cortisol. Cortisol is frequently referred to as a "stress hormone" because circulating levels increase dramatically in response to any stimulus that requires, or might require, metabolic resources. Thus, cortisol increases blood sugar levels, increases blood pressure, and reduces immune responses and inflammation (hence the use of cortisone cream or injections to control inflammation of the skin after exposure to poison ivy). This complex web of physiological links may seem far removed from feelings of social isolation, but loneliness has repeatedly been observed to be a risk factor for elevated levels of cortisol, especially in the morning. For instance, in middle-aged adults, the more intense the degree of loneliness reported at day's end over the course of three days in everyday life, the higher the spike in cortisol the subsequent morning. The conundrum is that loneliness is associated with increased risk of chronic conditions that are characterized by heightened inflammation (such as atherosclerosis, elevated cholesterol levels, heart disease, diabetes, and even cognitive impairment). If cortisol dampens inflammation, why might elevated levels of cortisol in lonely individuals be associated with more rather than less inflammation?

It turns out that communication among the hypothalamus, pituitary gland, and adrenal glands becomes disregulated when chronically stimulated. Whereas cortisol effectively dampens immune and inflammatory responses on an acute basis, when circulating cortisol levels are chronically elevated, cells become resistant to its immunosuppressant and antiinflammatory effects. This alteration happens at the level of DNA where the actions of genes in each cell of our body can be turned on (expressed) or off. Recent evidence suggests that the effects of loneliness reach down to this level. Circulating leukocytes (white blood cells) from a small group of chronically lonely adults showed decreased expression of glucocorticoid-response genes relative to expression rates in a matched group of socially connected adults. These genes are important because they activate the production of proteins that "hear" the anti-inflammatory signal sent by cortisol. Thus, despite higher levels of circulating cortisol in the lonely group, the cortisol signal may still not be heard. The lonely group also showed increased expression of genes carrying proinflammatory elements that, together with reduced expression of glucocorticoid-response genes, provides a functional genomic explanation for elevated risk of inflammatory disease in individuals who experience chronically high levels of loneliness.[9]

The effects of loneliness are not limited to an increase in health-threatening processes. Loneliness has also been associated with a decrease in health-restoring processes. Sleep is the quintessential example. Much of what feels stressful and depressing at the end of a long day is perceived differently following a good night's sleep. Good-quality sleep is the clincher. Lonely and nonlonely individuals do not differ in the amount of sleep they get, but lonely people have poorer sleep quality than do nonlonely people. They experience more microawakenings during the night, and they awake feeling more tired and less capable of meeting the demands of the day. Poor sleep has been associated with elevated blood pressure and cardiovascular mortality, and this may help to explain the poorer health outcomes of chronically lonely individuals. In short, lonely days extend into the nights and lessen the restorative nature of sleep.

Loneliness and health: It's not just peripheral

From this sampling of the widespread effects of loneliness on health, lifestyle behaviors, physiological functioning, genetic transcription, and sleep quality, it should be clear that the invisible power of felt isolation has long tentacles that have the potential to influence all physiology. Not only physical health, but also cognitive health is compromised. Indeed, one of the most sobering findings of recent years is that loneliness places people at risk for impaired cognition and dementia.[10] In a four-year prospective study of initially dementia-free older adults, the risk of Alzheimer's disease was more than twice as great in lonely than in nonlonely individuals. In addition, loneliness was associated with lower cognitive ability at baseline levels and with a more rapid decline in cognition during the four-year follow-up. Similar results were reported for a sample of 75- to 85-year-old individuals over a ten-year follow-up.

The effects of loneliness on cognition are evident at an even more fundamental level.[11] Self-regulation—the ability to regulate one's attention, cognition, emotion, or behavior to better meet social standards or personal goals—is impaired in lonely individuals. For instance, among young adults, instructions to shift auditory attention from the dominant right ear to the nondominant left ear resulted in impaired performance in lonely relative to nonlonely individuals. Loneliness also alters emotion regulation. In middle-aged and older adults, loneliness was associated with a diminished tendency to capitalize on positive emotions, and this explained why lonely individuals were less likely than nonlonely individuals to engage in physical activity. Impaired cognitive regulation is evident in experimental studies that manipulate feelings of isolation. Participants who receive feedback that induces a sense that they are doomed to a future of social isolation perform significantly worse on tests of reasoning, behave more aggressively, and choose more tasty but unhealthy foods than other participants who receive feedback indicating a future of social connection, or receive bad feedback of a nonsocial nature, namely that their future will consist of general misfortune.[12] There seems to be something uniquely threatening about the prospect, and the reality, of feeling isolated, disconnected, and like one doesn't belong.

In terms of emotional health, the prospect for lonely people is increasing misery, at least during the short term. Loneliness and depression tend to be thought of as synonymous, but the two are conceptually and empirically distinct. If loneliness and depression were synonymous, increases in loneliness would have no capacity to predict increases in depressive symptoms because increases in one would be exactly paralleled by increases in the other. Instead, longitudinal data have shown that loneliness predicts an increase in depressive symptoms, but depressive symptoms do not predict an increase in loneliness during a one-year interval.[13] Importantly, the influence of loneliness on depressive symptoms was not attributable to fewer social connections, general negativity, stress, or poor social support. These data suggest that the relevant intervention target is loneliness, and that modifying the cognitions, perceptions, and expectations of the lonely individual could help improve quality of life and overall well-being.

Social connectedness: invisible forces made visible

At this juncture, having become acquainted with the burden of loneliness, it is helpful to remember that most people, most of the time, feel socially connected. They derive satisfaction and meaning from their social relationships, and this makes them happier and more satisfied with life. Interestingly, happiness leads to higher levels of relationship satisfaction, indicating that happiness and relationship satisfaction feed forward to foster spirals of increasing happiness and relationship satisfaction. The general positivity that ensues from happiness is also apparent in the effects of happiness on income. Happiness predicted increases in household income during a two-year period in middle-aged adults. However, relationship satisfaction also predicted increases in household income during this time period and, remarkably, relationship satisfaction helped to explain the effect of happiness on income. It seems that happy people experience increases in income in part because of the general goodwill that surrounds the socially contented individual and that elicits tangible and intangible positivity from others.

It is perhaps precisely because most people feel socially connected and happy that we take for granted the invisible force of social connectedness and its stabilizing and nurturing influence in all aspects of life. Only in its absence do we begin to comprehend its power. Western notions of the autonomous individual notwithstanding, human beings are "wired" for social connections and need social bonds to feel safe, valued, motivated, and competent.[14] Among the lamentations expressed by some people in Western societies is a concern that our autonomy and independence come at the expense of meaningful social relationships and a sense of belonging to a larger social unit. Family members are no longer obligatorily part of our social community, while Facebook friends, some of whom we know only through electronic media, are deemed essential to fulfilling our need for a sense of connectedness and belonging.[15] The broadening of our social worlds has not been accompanied by maintenance, much less improvement, of the quality of our social relationships. One national study showed a threefold increase between 1985 and 2004 in the number of Americans who reported no one with whom to discuss important matters.[16] We are a meaning-making species. Relationships that offer security, comfort, trust, and pleasure, even if interactions are relatively infrequent, are much more effective at fostering a sense of connectedness and belonging than are more friends or more frequent interactions that fail to meet these standards. The challenge, especially for those of us who live in Western society, is to recognize that the invisible force of social connectedness has benefits for health and well-being that we ignore at our peril.

Conclusion

The research on loneliness highlights the need for and benefits of human connections, but it speaks even more directly to the role of beliefs about our connections. Loneliness, after all, is not about how many social relationships a person has; it's about a belief that the existing social relationships fail to satisfy a desired sense of social connectedness. All human relationships have a tangible existence in physical interactions and an invisible existence in mental representations and beliefs. This human capacity expands the range of possible relationships. For instance, humans form meaningful connections with pets, with television characters whom they have never met, and with deities who lack a material existence. We have seen the health impact of the invisible force of loneliness; do different kinds of invisible forces have different effects?

Endnotes

1. F. Buechner, *Telling the Truth: The Gospel as Tragedy, Comedy, and Fairy Tale* (New York: HarperCollins Publishers, 1977).

2. A. Caspi, H. Harrington, T. E. Moffitt, B. J. Milne, and R. Poulton, "Socially Isolated Children 20 Years Later," *Archives of Pediatric Adolescent Medicine* 160 (2006): 805–811.

3. H. Sugisawa, J. Liang, and X. Liu, "Social Networks, Social Support, and Mortality Among Older People in Japan," *Journal of Gerontology* 49 (1994): S3–13.

4. S. Cheng, "Loneliness-Distress and Physician Utilization in Well-Elderly Females," *Journal of Community Psychology* 20 (1992): 43–56; J. Geller, P. Janson, E. McGovern, and A. Valdini, "Loneliness as a Predictor of Hospital Emergency Department Use," *The Journal of Family Practice* 48 (1999): 801–804.

5. L. C. Hawkley, and J. T. Cacioppo, "Aging and Loneliness: Downhill Quickly?" *Current Directions in Psychological Science* 16 (2007): 187–191.

6. L. C. Hawkley, M. E. Hughes, L. J. Waite, C. M. Masi, R. A. Thisted, and J. T. Cacioppo, "From Social Structural Factors to Perceptions of Relationship Quality and Loneliness: The Chicago Health, Aging, and Social Relations Study," *Journal of Gerontology: Social Sciences* 63B (2008): S375–S384.

7. J. T. Cacioppo and B. Patrick, *Loneliness: Human Nature and the Need for Social Connection* (New York: W. W. Norton & Company, 2008).

8. L. C. Hawkley, R. A. Thisted, C. M. Masi, and J. T. Cacioppo, "Loneliness Predicts Increased Blood Pressure: Five-Year Cross-Lagged Analyses in Middle-Aged and Older Adults" (under review).

9. S. W. Cole, L. C. Hawkley, J. M. Arevalo, C. Y, Sung, R. M. Rose, and J. T. Cacioppo, "Social Regulation of Gene Expression in Humans: Glucocorticoid Resistance in the Leukocyte Transcriptome," *Genome Biology* 8 (2007): R189.1–R189.13.

10. R. S. Wilson, K. R. Krueger, S. E. Arnold, J. A. Schneider, J. F. Kelly, L. L. Barnes, et al., "Loneliness and Risk of Alzheimer's Disease," *Archives of General Psychiatry* 64 (2007): 234–240.

11. J. T. Cacioppo and L. C. Hawkley, "Perceived Social Isolation and Cognition," *Trends in Cognitive Science* 13 (2009): 447–454.

12. R. F. Baumeister, C. N. DeWall, N. J. Ciarocco, and J. M. Twenge, "Social Exclusion Impairs Self-Regulation," *Journal of Personality & Social Psychology* 88 (2005): 589–604; J. M. Twenge, R. F. Baumeister, D. M. Tice, and T. S. Stucke, "If You Can't Join Them, Beat Them: Effects of Social Exclusion on Aggressive Behavior," *Journal of Personality & Social Psychology* 81 (2001): 1058–1069.

13. J. T. Cacioppo, L. C. Hawkley, and R. Thisted, "Perceived Social Isolation Makes Me Sad: Five-Year Cross-Lagged Analyses of Loneliness and Depressive Symptomatology in the Chicago Health, Aging, and Social Relations Study," *Psychology and Aging* (2009, in press).

14. R. I. M. Dunbar and S. Shultz, "Evolution in the Social Brain," *Science* 317 (2007): 1344–1347.

15. L. Pappano, *The Connection Gap: Why Americans Feel So Alone* (Piscataway, N.J.: Rutgers University Press, 2001).

16. M. McPherson, L. Smith-Lovin, and M. T. Brashears, "Social Isolation in America: Changes in Core Discussion Networks over Two Decades," *American Sociological Review* 71 (2006): 353–375.

From relationships to people and groups to relationships with God

The extent and quality of our social connections can have profound consequences for our physical well-being. In the previous chapter, Louise Hawkley explores, in particular, the consequences that feelings of inadequate social connection have on such physical outcomes as sleep quality, high blood pressure, reduced ability to respond to inflammation, cognitive health in aging, and cardiovascular health. While Hawkley emphasizes the relationship between the invisible forces of social connection and health, and the biological mechanisms responsible for this relationship, Gary Berntson takes things one step further by exploring a person's perceived connection with God and its effects on our basic biological systems. Many of our basic biological processes, such as breathing or maintaining sufficient blood pressure to oxygenate the brain, are reflexes that are so automatic that they become invisible to us. Berntson shows that thoughts and beliefs alter not only behaviors, but also the regulation of these reflexlike mechanisms. And he suggests that the root of these effects may lie in the way humans and other animals maintain biological equilibrium with regulatory mechanisms that, under ordinary circumstances, keep each other in check. This theory describes a biological mechanism that could explain why spirituality is associated with generally better health.

5*

Psychosomatic relations: from superstition to mortality

People have many sources of information, knowledge, and understanding. We consider the most common of these to be empirically acquired—learned facts, relations, associations, and perceptual and motor skills. Such learned associations serve as powerful determinants of thought and behavior. But other sources of information and knowledge also affect our

* The lead author is Gary G. Berntson, Ph.D., Professor of Psychology, Psychiatry, and Pediatrics, and a member of the Neurosciences Graduate Faculty at the Ohio State University. He is a cofounder (with John Cacioppo) of the field of social neuroscience, a coeditor of the *Handbook of Psychophysiology* and the *Handbook of Neuroscience for the Behavioral Sciences,* and the President of the Society for Psychophysiological Research. Berntson's research focuses on the evolutionary development of the neuraxis, with special regard to levels of organization in

interaction with the environment, including reflex-like (constitutionally endowed) circuits that are independent of explicit learning. Examples include central networks for pain withdrawal, hunger circuits for the ingestion of essential nutrients, social affiliation networks, and neural systems that promote maternal bonding. Each of these sources of information or knowledge can impact thoughts and beliefs, and thoughts and beliefs can impact behaviors and other bodily functions.

> [A] Maori woman who, having eaten some fruit, was told that it had been taken from a tabooed place; she exclaimed that the sanctity of the chief had been profaned and that his spirit would kill her…. The next day…she was dead.[1]
>
> I have seen a strong young man die…the same day he was tapued (tabooed); the victims die under it as though their strength ran out as water….[1]

A superstition is a belief based not on reason or knowledge, but on legend, magical thinking, or other nonrational basis. Beliefs color the way we perceive the world; they direct and shape our actions and define our personalities. Beliefs are powerful determinants of action. A useful illustration of the power of beliefs comes from the parable of the Sultan (who had studied psychology) and his "lie-detecting" donkey. Lore has it that the Sultan was missing a valuable vase from his estate and suspected that one of his servants had stolen the piece. To identify the culprit, the Sultan gathered his servants in front of a dark room in which a donkey was tied and then asked each of his servants if they had stolen the item. Each said "No." The Sultan explained that inside the room was a magical

---

neurobehavioral systems, affective processes, and autonomic regulation. He has published more than 200 articles in scientific outlets and 6 books.

Berntson begins with the fact that knowledge, thoughts, and beliefs can influence our behaviors. Behaviors, of course, are physiological processes entailing neural operations and muscular actions. Here we see a clear intersection between the psychological domain (knowledge, thoughts, and beliefs) and the physical domain (neuromuscular effector systems). But mind–body relations extend beyond the observable actions of skeletal muscles. The mind and its organ, the brain, also impact powerfully on internal bodily functions associated with the autonomic nervous system, the endocrine system, and the immune system, to name just a few. Through these interactions, psychological processes can be translated in outcomes that have powerful significance for adaptation and health. Berntson explores the processes by which thoughts can manifest in fundamental changes in internal physiology and health.

donkey, specially trained to detect liars, who would bray when slapped by someone who had lied. The servants were sent into the room, one by one, and were instructed to close the door, slap the donkey, and return. "When the donkey brays," the Sultan loudly proclaimed, "I will have my culprit." The first servant was sent into the room and returned shortly thereafter—tremendously relieved, as the donkey had not brayed. One by one, the remaining servants entered the room and returned. The donkey had not brayed and all the servants looked quite relaxed. The Sultan was sanguine—he knew this donkey never brayed under any circumstances. The Sultan asked the servants to show him their hands. He then pointed to one of them and declared, "We have our thief," instructing the guards to take him away. How had he identified the culprit? Rather than relying on a magical donkey, the Sultan, who was a student of psychology, took a more rational approach. Understanding the impact of beliefs on behavior, the Sultan had surreptitiously infiltrated powdered charcoal into the donkey's hair. When the servants slapped the donkey, the charcoal marked their hands—with the exception of the guilty servant, who had not slapped the donkey, out of a belief and associated fear that the donkey could detect a liar.

Power of beliefs

Beliefs may be potent determinants of behavior, but can they kill? And if so, how? How can these invisible, intangible entities impact health? In a now classic article published in the *American Anthropologist* in 1942, Walter Cannon, a leading Harvard physiologist and expert on the autonomic nervous system, proposed an answer.[1] Investigating phenomena such as voodoo practices of the Haitians and "bone-pointing" among Australian aborigines, Cannon found that a common feature among the victims of such rituals was a strong belief in the curse and an associated morbid fear of the outcome. That fear, he argued, could trigger a "fight-or-flight reaction" (a phrase he had earlier coined), characterized by powerful and exaggerated activation of the sympathetic nervous system. The resulting vascular constriction diminishes blood flow to critical tissues (ischemia), with consequent hypoxia (decreased oxygen) and disturbances in normal metabolism and cellular function. These reactions may be exacerbated by the lack of food and water as the victim "pines away." Cannon argued that these reactions could become life-threatening—fulfilling the gruesome legacy of the ritual—based on a belief in the supernatural, the veracity of which is largely irrelevant. More relevant is the emotion triggered by the belief, specious as it may be.

Beliefs and emotions have consequences, both behavioral and physiological. A recent example comes from the contemporary medical literature. There is now a well-documented condition, sometimes triggered by something as innocuous as a spousal argument or a surprise birthday party, that entails the hallmark clinical and physiological features of a heart attack, including chest pain, abnormalities on the electrocardiogram, and elevated cardiac enzymes (reflecting damaged heart muscle).[2] The condition has variously been termed *takotsubo cardiomyopathy*, *left-ventricular apical ballooning, myocardial stunning, stress cardiomyopathy,* or, in the more vernacular parlance of *The New York Times, broken heart syndrome* (prompted by a medical review that was published just before Valentine's Day). In general accord with the speculation of Cannon, broken heart syndrome appears to be triggered by an exaggerated autonomic nervous system response, characterized by sympathetic activation and high levels of the stress hormone epinephrine (adrenalin).[3] It is important to note in these cases that psychological states, as mild as they may be, are able to induce a clear and demonstrable organ pathology.

Physiological abnormalities or dysfunctions underlie medical conditions and, indeed, constitute the defining features of disease states. An important question, however, is how those dysfunctions come to be. Disease develops in many ways—traumatic injuries, biotic infections, degenerative conditions, and the list goes on. The fields of psychophysiology, psychosomatic or behavioral medicine, and health psychology are particularly concerned with how psychological and behavioral factors impact physiological systems and, therefore, health. Of particular interest are those psychological dimensions that uniquely impact physiology.

An example comes from the study of Herpes Simplex viral infections. Herpes Simplex viruses (HSV) are responsible for cold sores (HSV type I) and genital herpes (HSV type II). Once contracted, herpes virus infections generally remain for life, although they are characterized by periodic eruptions and remissions. During the latter, the immune system effectively dampens viral activity, and the virus retreats to an essentially dormant state. Although multiple factors likely contribute to the reactivation of HSV, one trigger appears to be stress—the defacing cold sore that erupts, for example, just before the prom or an important date. Ohio

State researchers sought an animal model of this reactivation so the underlying links and mediators could be studied. Try as they might, however, the research group was unable to reactivate HSV infections in mice with standard laboratory stressors such as restraint stress or shock. In a collaborative effort, we pointed out that the stressors that lead to HSV reactivation in humans were often of a social nature. Indeed, for both humans and mice, social relations are central to happiness, adaptation, and even survival. In light of this, a social stressor was introduced into the project (changing the housing groupings and thus disrupting established social relations). The social stress, not physical stressors, resulted in significant HSV reactivation.[4] Psychological factors, in this case a specific social psychological variable, uniquely impacted an important aspect of viral immunity. This early finding led to a series of studies that have elucidated physiological pathways that mediate the relationship among social stress, immune function, and HSV reactivation. But what is it that makes social stress unique and distinct from physical stressors?

We have identified a probable general contributor to the differences between lower-level physical or homeostatic challenges and higher-level psychological and social stressors. Basic homeostatic reflexes—reflexes that keep in balance various critical bodily processes, such as blood pressure, body temperature, and blood sugar—are largely hard-wired and organized at relatively low levels of the nervous system, such as the brainstem and spinal cord. An example comes from autonomic nervous system regulation of cardiovascular function. The sympathetic division of the autonomic nervous system is an activational, energy-mobilization system that comes into play in the face of adaptive challenges. Sympathetic activation increases heart rate and results in peripheral vasoconstriction, both of which tend to increase blood pressure. In contrast, the parasympathetic division is an energy-conserving, deactivational brake that generally opposes the sympathetic system, yielding decreases in heart rate and blood pressure. The baroreceptor heart rate reflex is a homeostatic reflex that functions to maintain blood pressure within homeostatic limits. Unique pressure-sensitive receptors in the heart and large arteries detect changes in blood pressure, and a decrease in blood pressure triggers the baroreceptor heart rate reflex, increasing sympathetic activity and reciprocally decreasing parasympathetic tone. Both effects serve to increase heart rate (and thus cardiac output) and constrict arteries throughout the body, thereby restoring the pressure perturbation. In basic reflexes, the two autonomic branches are generally regulated in this reciprocal fashion and thus synergistically amplify the effects of the other. This is a useful mechanism to adjust to severe adaptive challenges, such as a decrease in blood pressure and compromised circulation.

Although this reciprocal mode of regulation of the autonomic branches has considerable utility and is characteristic of basic reflex organizations, it may not always be optimal. The autonomic nervous system provides the basic support for action and adjustment, and although it figures prominently in survival-related functions, it also provides the basic visceral support for emotional and cognitive operations. It has long been recognized that cognitively demanding tasks elicit greater autonomic activation than is needed to meet the metabolic demands of the tasks. Moreover, ascending neural signals to the brain from visceral organs such as the heart and blood vessels serve to modulate and regulate cognitive activities.[5] The notable early psychologist William James proposed that emotion is the experience of somatovisceral sensory feedback. James suggested that we do not run from the bear because we are afraid, but rather we are afraid because we run from the bear.[6] Although the strong form of this theory has not been supported, it remains the case that ascending visceral signals can modulate learning, attention, and cortical and cognitive processing.[5] The autonomic nervous system is not only for lower-level reflexive adjustments. Indeed, it is increasingly recognized that there is a highly complex, even intricate, interaction between the autonomic nervous system and higher-level brain structures (such as the frontal cortex) involved in human behavior. Importantly, these circuits and their interactions with the autonomic nervous system are highly flexible and are not constrained by the simple organization rules that govern basal functions such as homeostasis and reciprocal control of the autonomic branches. Rather, higher-level systems engage in highly sophisticated "banter" with the autonomic nervous system.

In contrast to the reciprocal control characteristic of autonomic reflexes, higher-level brain circuits exert more flexible control over the autonomic nervous system. This can include the classic reciprocal control pattern but can also include an independent control pattern in which only the sympathetic branch or only the parasympathetic branch of the autonomic nervous system is activated, and a coactive control pattern in which both branches are activated. This greater flexibility in control may have behavioral and health significance.

Beliefs about one's relationship with God and autonomic functioning

Recently, we used a population-based sample of 50- to 68-year-old adults in the Chicago Health, Aging, and Social Relations Study to examine risk factors for heart attacks—a health outcome known to be influenced the autonomic nervous system. We were particularly interested in whether spirituality influenced risk for heart attacks. With very few exceptions, everyone in our sample expressed a belief in God. However, individuals differed in how they perceived the quality of their relationship with God, much as individuals differ in how they perceive the quality of their relationships with other people. We defined spirituality as the degree to which a personal relationship with God was believed to offer safety, security, contentment, and love. One observation that emerged from this study was that spirituality was associated with a lower incidence of heart attacks.[7] This remained true after ruling out the effects of demographics, health behaviors, body mass index, blood pressure, and other potential explanatory factors. Short of divine intervention, was there a rational explanation for this relationship? We certainly know that psychological factors can impact autonomic control, among other aspects of physiology.

As mentioned earlier, sympathetic activation may have harmful consequences if it is extreme or prolonged. For example, heightened sympathetic activation is known to predict a poorer outcome after a heart attack. In contrast, parasympathetic activity may have beneficial or protective effects. From the perspective of a reciprocal model of autonomic control, high parasympathetic with low sympathetic control would be optimal, whereas high sympathetic with low parasympathetic control would be considered a risk. But we also know that higher-level neurobehavioral systems may not be constrained to reciprocal autonomic controls. Moreover, it has been argued that more autonomic control is better than little control, in that it affords greater capacity for adjustment of visceral functions. Could high levels of parasympathetic control, for example, mitigate the negative effects of sympathetic activation, and perhaps yield an even more advantageous health outcome?

To examine these questions, we developed two quantitative measures of autonomic control.[8] The first was a common metric of autonomic balance (Cardiac Autonomic Balance), which represents the relative dominance of the two branches along a single autonomic continuum that ranges from purely parasympathetic control to purely sympathetic control. This metric is consistent with the classical model of reciprocal control, characteristic of reflex processes, in which autonomic balance can be biased toward one of the autonomic branches. High scores indicate sympathetic dominance and low scores indicate parasympathetic dominance. Independent estimates of sympathetic and parasympathetic control were obtained using standard measurement procedures, and the level of parasympathetic control was subtracted from sympathetic control to derive a measure of Cardiac Autonomic Balance. A second metric was designed to capture an alternative mode of autonomic control (Cardiac Autonomic Regulation) that assesses the degree of relative coactivation (rather than reciprocal activation) of both branches. This is a metric that taps into the nonreciprocal regulatory influences of higher neural structures. Cardiac Autonomic Regulation scores were derived by essentially summing activities of the sympathetic and parasympathetic branches to afford a measure of total overall autonomic cardiac control. High scores indicate high activation and low scores indicate low activation of both branches of the autonomic nervous system.

In the 50- to 68-year-old adults in our sample, spirituality was found to be associated not only with a lower incidence of a heart attack, but also with a higher level of Cardiac Autonomic Regulation. That is, people who felt a closer relationship with God exhibited higher overall autonomic regulation—both sympathetic and parasympathetic. This was associated, in part, with lesser diminution of parasympathetic control and a greater degree of coactivation. Moreover, Cardiac Autonomic Regulation (but not Cardiac Autonomic Balance) predicted better overall health status and was associated with a lower incidence of a heart attack. Participants who had low Cardiac Autonomic Regulation were more likely to have suffered from a heart attack.

Could higher Cardiac Autonomic Regulation scores explain why spirituality was associated with less risk for a heart attack? That is, could a pattern of autonomic regulation associated with spirituality explain the link between spirituality and a heart attack? To address this question, we conducted statistical tests of these linkages. As we already knew, both spirituality and Cardiac Autonomic Regulation are associated with a lower incidence of a heart attack. When the predictive effects of spirituality were statistically extracted, Cardiac Autonomic Regulation continued to be a significant predictor of a lower incidence of heart attacks. However, when the linkage test was reversed and the effects of Cardiac Autonomic Regulation were extracted, spirituality was no longer a significant predictor. This indicates that Cardiac Autonomic Regulation is a plausible mediator that may explain the relationship between spirituality and heart attacks.

By capturing higher levels of parasympathetic control and the associated autonomic coactivation of the sympathetic and parasympathetic branches, Cardiac Autonomic Regulation provided a critical metric that permitted the study of a previously "invisible" and mysterious link between spirituality and health outcomes. This in no way diminishes the relationship between spirituality and health, but it instead offers an important hypothesis on how spirituality may impact physiology and health status. Spirituality reflects an important aspect of the general domain of sociality and social relationships—a domain heavily influenced by our genetic constitution as a social species. Indeed, the importance of sociality may be more related to beliefs and attitudes about the meaningfulness of relationships than their existence or number. And again, beliefs about social relationships also have real consequences.

Conclusion

Beliefs impact thoughts and actions. This may be reflected in phenomena as diverse as biasing a behavioral disposition (such as slapping a donkey), coloring our perception of the environment, or determining how we perceive the quality of our social (including spiritual) relations. Psychology can impact physiology, and physiology, in turn, can influence our thoughts and emotions. Psychophysiology is the study of these relationships, and it promises to illuminate the intricacies of psychosomatic relations and the previously "invisible" mechanisms that mediate these links. The relations between the mind and the body, the so-called mind–body problem, are complex and still rather obscure. Nevertheless, the mind–body problem is yielding to science, and the *problem* it poses is progressively diminishing. Among the components of the mind–body problem yielding to rigorous scientific inquiry are the effects of spiritual beliefs, including feelings of closeness to God.

Endnotes

1. W. B. Cannon, "'Voodoo' Death," *American Anthropologist* 44 (1942): 169–181.

2. I. S. Wittstein, "The Broken Heart Syndrome," *Cleveland Clinic Journal of Medicine* 74, Supplement 1 (2007): S17–22.

3. I. S. Wittstein, D. R. Thiemann, J. A. Lima, K. L. Baughman, S. P. Schulman, G. Gerstenblith, K. C. Wu, J. J. Rade, T. J. Bivalacqua, and H. C. Champion, "Neurohumoral Features of Myocardial Stunning Due to Sudden Emotional Stress," *New England Journal of Medicine* 352 (2005): 539–548.

4. D. A. Padgett, J. F. Sheridan, J. Dorne, G. G. Berntson, J. Candelora, and R. Glaser, "Social Stress and the Reactivation of Latent Herpes Simplex Virus—Type I," *Proceedings of the National Academy of Sciences* 95 (1998): 7231–7235.

5. G. G. Berntson, M. Sarter, and J. T. Cacioppo, "Ascending Visceral Regulation of Cortical Affective Information Processing," *European Journal of Neuroscience* 18 (2003): 2103–2109.

6. W. James, "What Is an Emotion?" *Mind* 9 (1884): 188–205.

7. G. G. Berntson, G. Norman, L. Hawkley, and J. T. Cacioppo, "Spirituality and Autonomic Cardiovascular Control," *Annals of Behavioral Medicine* 35 (2008): 198–208.

8. G. G. Berntson, G. J. Norman, L. C. Hawkley, and J. T. Cacioppo, "Cardiac Autonomic Balance vs. Cardiac Regulatory Capacity" *Psychophysiology* 45 (2008): 643–652.

The mind and body are one

The Cartesian view of the mind as distinct from the body persists in twenty-first century discourse as the mind-body problem alluded to by Gary Berntson. Berntson provides evidence that the mind and the body, psychology and physiology, are not independent of each other, but represent different levels of functional organization of human organisms. Beliefs influence thoughts, behaviors, and physiology, and peripheral physiological processes signal central neural networks that influence cognitions and feelings crucial for the generation and moderation of beliefs. Spiritual beliefs are considered by some to be contentious candidates for scientific examination, yet Berntson argues that spiritual beliefs can be identified, measured, and subjected to scientific investigation in the same fashion as any other belief or invisible force. Accordingly, Berntson examines the effects of a specific belief—the belief that one has a solid spiritual life. As documented by Berntson, this belief may be associated with rather profound physiological and health effects.

Whereas Berntson focuses on the influence of the mind *on* the body and vice versa, Gün Semin speaks of the mind *in* the body and, more specifically, in several bodies simultaneously. In his social cognition model, Semin challenges the limits of individual social cognition and argues that regulation and coregulation of social behavior are distributed across brains. When several individuals exhibit spontaneous synchronized behaviors (such as hand-clapping), the human tendency is to invoke a "supraindividual" explanation. Semin describes a mechanism by which the supraindividual source can be explained as shared motor representations and ongoing monitoring of observed actions that, under certain conditions, lead to dissolution of the boundary between self and other. The resulting shared experience of unity and collective identity may feel transcendental, but the mechanisms are as real and explicable as those governing individual behaviors and experiences.

6 *

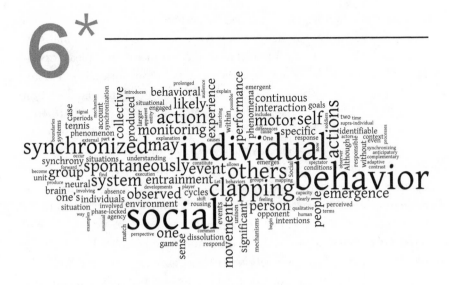

The suspension of individual consciousness and the dissolution of self and other boundaries

When we watch a group of soldiers marching in formation, we see the behavior of the group synchronized. Although we can make out the individual within the group, the group seems to be an entity of its own and the individual soldier seems to have become a cog in the social machine. Mob behavior, crowds at sporting events, and soldiers in formation all

* The lead author is Gün R. Semin, Ph.D., an Academy Professor, Royal Nether-lands Academy of Arts and Sciences, at Utrecht University, The Netherlands. He is the founding Scientific Director of the Kurt Lewin Graduate School, a past president of the European Association of Experimental Social Psychology, and the Chair of the International Committee of the Association for Psychological Science. Semin's research is primarily driven by an interest in communication, social cognition as a jointly recruited process, and language and the diverse uses

73

suggest that when we are organized to act together, the group becomes an emergent entity that can submerge the sense of the individual self. This apparent social absorption stands in contrast to our typical experience of being autonomous, self-aware agents in the world. The dissolution of the boundary between the self and the group is one manifestation of the social brain and the mechanisms that support our ability to connect with others.

Although not everyone has the experience of marching in a band or running with a mob, most people have been part of an audience at a concert or play. At the end of a particularly thrilling performance, an audience can be moved spontaneously as a group to clap wildly. In these situations, we know the feeling of surging to our feet as a collective, hands clapping and faces beaming with approval. As the clapping blooms, individual clappers merge into a synchronized unit.[1, 2] Similarly, thousands of individual sports fans have been observed to stand and raise their hands in a synchronized fashion to produce a collective wave that travels around the stadium. In both cases, individual people act as a collective unit, a superorganismal structure, with capacities and behaviors beyond the reach of any single individual in the group.

These examples are instances of behavioral uniformity in large groups. What is distinctive about these examples is that the observed behavioral synchrony can be understood in terms of a shared social goal. Sometimes the goal is spontaneous, as in clapping to demonstrate appreciation, and sometimes the goal is imposed by the situation (such as musicians following a musical score and instructions of a conductor, or soldiers marching to the call of a drill sergeant). However, when we

---

that language can be used for in social interaction (ranging from the regulation of prejudice to interpersonal relationships), as well as the embodied grounding of meaning and communication.

A puzzle that has occupied Semin much of his career is how it is possible to understand social behavior by explaining individual processes. Another puzzle has been why one should focus on stills when human behavior is a movie: Behavior is self-evidently dynamic and highly responsive to contextual variations. Finally, Semin has been puzzled by how it is possible to think that psychological processes might be only some symbolic computation taking place somewhere between the ears. As outlined in this chapter, he has come to conceptualize the social in terms of jointly recruited processes rather than individual ones, social behavior as situated, and psychological processes as embodied.

behave in synchrony with others, there is a sense of becoming part of something larger than ourselves.

To the people engaged in spontaneously synchronized behavior, there is a clearly identifiable and seemingly individual "cause" for their emergent behavior. But when a group shares the same goal—demonstrating approval—and engages in the same action—clapping—the stage is set for such behavior to become coordinated and organized even without an external agent (such as a conductor or drill sergeant). How do those moments of spontaneous social aggregation occur? How does the social brain work to join with others to form the emergent group?

We have begun to understand the underlying dynamics of how and when such phenomena are likely to occur, and even how such phenomena can be potentially engineered. New insights afforded by developments in social psychology, developmental psychology, and social neuroscience have suggested the way in which our brains respond to the invisible force of social connection. These scientific developments suggest neural mechanisms that may be important to the way we interact with others. At the same time, the insights also reveal the likely conditions under which the individual self merges into the group. Such situations when the sense of self is suspended contrast sharply with the modern Western notion of the individual standing apart from others. Indeed, the traditional Western focus on individual-centered reasons, motives, intentions, and causes may be at odds with some forms of spontaneously synchronized behaviors and group action.

Toward a biology of social interaction

Consider the perspective of an engaged spectator at a singles tennis match. Although we may be sitting distant from competitors, if we identify with one of the players, we are not merely passive observers. On the contrary, our observation of the events in the game can serve to activate some of the same neural mechanisms that would be active if we were playing the game rather than just observing it. We can feel the moves, the impetus to defend an attack, and the urge to slam the ball as if we ourselves are playing, albeit without actually flailing our arms around. We may even anticipate a move by the opponent and imagine ourselves making the potential response. Research during the last ten years has revealed that our brains can map the movements of other human beings onto our own bodies, almost as if we were making those movements.[3] This capability to put ourselves in another person's shoes makes it possible to identify with either player.

By comparison to the audience, consider how this capability can serve us as one of the players. This capacity provides an important facility for anticipating our opponent's moves, allowing us to plan a response even before the opponent has completed a ground stroke. This kind of anticipation does not depend on explicit reasoning or conscious reflection—it seems to operate as an automatic mechanism.[3, 4] This kind of mechanism may facilitate understanding the behavior of others. If your brain mirrors the neural activity in the brain of someone you see acting, this could provide a basis for understanding the motivation for the action. If your brain resonates to the observed action as if you were acting, this could call to mind previous experiences acting that way and provide a memory for why you acted that way. That is, our social brain may directly resonate to the actions of others without reasoning explicitly about those actions. This kind of mechanism, through which intentions might be inferred, could then prepare responses quickly to facilitate the smooth flow of social interactions, whether in a game or a dialogue. Of course, a critical aspect of such a mechanism is to differentiate our resonance to other people's actions and the control of our own. This kind of neural system for mapping the actions and intentions of others has been identified with a network of regions called the mirror neuron system,[5] and this system may help induce a degree of reflexive similarity or identification between self and other. The mirror neuron system appears to be continuously engaged, unless it is actively suppressed by inhibition, so that it may continuously monitor the behavior of "others" in our social environment.

Of course, mapping the movements of the opponent is useful, but certainly not sufficient to defend our position, score a point, or win a match. One needs to execute countermoves. This is the domain of the motor system in the brain, which includes regions involved in the preparation and execution of motor action. The motor system is responsible for the implementation of one's goals and intentions to perform an action.[6] Thus, the social brain includes monitoring and motor systems that function in parallel. The mirror system puts the player in the opponent's shoes and monitors the opponent's actions in an anticipatory manner. Other parts of the social brain maintain the distinction between player and opponent by shaping the implementation of one's actions, namely by engaging a counteraction.

In sum, a tennis game or any social interaction depends on a complex network of brain regions that mediate perception and action, and the relationship between observed action and one's own behavior. The overlap in brain regions responsible for these two important social functions suggests how tightly coupled and coordinated social interactions can be. However, these two systems cannot operate in isolation from our knowledge of the context in which behavior occurs. Therefore, we turn to this topic next.

The social context

In a tennis game or any social interaction (such as dancing or conversation), the behavior of one individual constitutes a stimulus for others. If a behavior is meaningful, neural mechanisms responsible for social perception and social interaction are likely to be activated to engage in complementary action. In a competitive context, such as the tennis game, the motor system is engaged in the preparation and execution of complementary actions to those observed and anticipated based on the inferred goals and intentions of the competitor. However, if everyone shares the same goal, such as an audience clapping, the neural systems for monitoring the actions of others and executing one's own actions can be mutually reinforcing, leading to synchrony.

In any dynamic social situation, observing one person's action can initiate neural activity in another. Parts of the observer's motor system become activated, making it possible to respond in synchrony. In fact, the specific actions that are observed are not as important as the perceived goals or intentions of the observed person. One consequence of this is that a person is sensitive to new actions in the social environment. A second consequence is that significant actions by another person can quickly produce complementary motor responses. Thus, adaptive social behavior is a product of perceptual monitoring and motor processes. The complexity of the social environment necessitates *selective responding to socially significant features of any social interaction*. Such selective responding depends on the observer having particular goals for action. The identification of significant stimuli (such as a threatening backhand smash in tennis) activates in the competitor's brain goal-driven decision processes that operate in parallel with continuous social monitoring and lead to a counteraction (such as a defensive lob) produced by the motor system. However, a linesman collecting a ball during a tennis match does not constitute a significant action for the competition and, even if observed, does not initiate any counteraction.

Let us now return to the perspective of a spectator at the tennis match. Although the specific movements driving the tennis match have significant implications for the players' actions, these movements have a different implication for the spectator, from whom no overt responses are warranted. If an observed action produced by someone else does not have personal significance for an observer, the motor system does not respond in the same way.[7] The goal-dependent aspect of observing the actions of others allows us to understand and respond quickly and effectively without confusing what we do with what we see. However, in some special cases, such as when a group of individuals all respond together, the same motor system may operate differently. In these situations, we are neither observer nor respondent, but part of a flock or chorus that our sense of individuated self may begin to dissolve into the larger social group.

The social parameters for suspended self-consciousness

The dissolution of self–other boundaries is likely to be manifest under a specific set of conditions, which includes a strong feeling of identification with (connection between) oneself and a group of others, the absence of constraints to action by oneself or the observed others, a common goal shared by the group, and the absence of a recognized external synchronizing signal to which one can attribute any synchronized behavior. Clapping in unison following a rousing performance is a more common example. This remarkable phenomenon is evidenced despite considerable individual differences in clapping tempos. The transition to entrained clapping, whereby each clapper affects the surrounding other clappers both locally and globally, enhances the noise intensity at the moment of the clapping even though it leads to a decrease in the overall average noise intensity in the room. Synchronized behavior occurs rhythmically, and one way of capturing its regularities is to model its cycles, periods, frequencies, and amplitudes.[1, 2] Depending on the particular behavior and interaction in question, behavioral cycles of interpersonal entrainment can range from milliseconds to hours. Indeed, this kind of interpersonal entrainment is a pervasive phenomenon not specific to human social behavior.[8]

When does synchronized clapping occur? The distinctive feature of such an event is a convergence between the neural mechanisms underlying the monitoring of the movement of others and the execution of one's own movements. The specific factors responsible for the tipping point from asynchronous to synchronous clapping are not yet known, but descriptively each individual shifts the timing of his or her subsequent clap to the perceived timing of claps by the whole collective. Thus, a continuous adjustment process emerges in the form of a collective behavior (synchronized clapping) to which each individual contributes and no single individual controls. Such continuous monitoring of the collective rather than individuals within the collective and the adjustment of one's own movements to synchronize one's behavior with that of the collective result in a continuous loop of performing the very same action, leading to dissolution between self and other and the emergence of an entrained unit. The resulting effect is the materialization of a *supraindividual* behavioral phenomenon, namely extended behavioral cycles that are locked together in time. Although the emergence of clapping in unison can be regarded as a phenomenon worthy of more detailed understanding, it does not induce in its performers the necessity of searching for an explanation, since the readily available account is that the performance somehow produces synchronous clapping.

Preliminary evidence suggests that even with no externally imposed demands, prolonged synchronization emerges within pairs of people interacting, or *dyads*.[9] Despite individual differences in movements, participants entrain (tap together in time) rapidly when participants can perceive the behavior of others.[9] Extensive research in coordination dynamics has demonstrated that such entrainment does not depend on the intention to coordinate behavior.[10, 11] Studies have repeatedly shown that there is a spontaneous propensity to mimic other people (generally observed in dyads). One implication of this kind of behavioral synchrony is the emergence of affective bonding—such as a feeling of rapport—both in the case of *mimicry*[12] and between the synchronized partners.[13] One might conjecture that positive feelings are even stronger when more people are synchronized.

The conditions that lead to behavioral synchrony can vary. It is interesting that people may be more likely to experience the dissolution of self–other boundaries when synchronization is produced without any obvious external director and is continuous. Prolonged spontaneous and unintended entrainment among three or more people may introduce this feeling, because the emergence of synchrony cannot be attributed to a single external cause.

Conclusion

Social interaction involving extended periods of synchronized behavior is not part of our daily experience, particularly when it involves more than two people. How do we understand this kind of synchrony when it occurs? In the Western intellectual tradition, we have a strong tendency to search and explain events in terms of individual agency and causation.

Social events are generally understood in terms of contributions of the individual and the situation itself. The degree to which the person or the situational constraints shape the nature of the event will vary greatly. However, spontaneous and prolonged entrainment among three or more people introduces an experience that is difficult to explain by these more traditional accounts. These experiences cannot be easily reduced to the actions of a single person, so an account has to be found in some source that goes beyond the individual. The powerful sense of unity and belonging that emerges from this kind of experience almost demands a different kind of explanation than we generally consider. Indeed, such feelings emerging from the synchrony of behavior may provide some of the foundation for the cultural interpretation of a transcendental experience.

Endnotes

1. Z. Néda, E. Ravasz, Y. Brechet, T. Vicsek, and A. L. Barabasi, "The Sound of Many Hands Clapping: Tumultuous Applause Can Transform Itself into Waves of Synchronized Clapping," *Nature* 403 (2000) : 849–850.

2. Z. Néda, E. Ravasz, T. Vicsek, Y. Brechet, and A. L. Barabasi, "Physics of the Rhythmic Applause," *Physical Review* 61 (2000): 6987–6992.

3. G. R. Semin and J. T. Cacioppo, "Grounding Social Cognition: Synchronization, Entrainment, and Coordination," in G. R. Semin and E. R. Smith (eds.), *Embodied Grounding: Social, Cognitive, Affective, and Neuroscientific Approaches* (New York: Cambridge University Press, 2008).

4. G. R. Semin and J. T. Cacioppo, "From Embodied Representation to CoRegulation," in J. A. Pineda (ed.). *Mirror Neuron Systems: The Role of Mirroring Processes in Social Cognition* (Totowa, New Jersey: Humana Press, 2009).

5. M. Iacoboni, R. P. Woods, M. Brass, H. Bekkering, J. C. Mazziotta, and G. Rizzolatti, "Cortical Mechanisms of Human Imitation," *Science* 286 (1999): 2526–2528.

6. C. L. Elsinger, D. L. Harrington, and S. M. Rao,. "From Preparation to Online Control: Reappraisal of Neural Circuitry Mediating Internally Generated and Externally Guided Actions," *NeuroImage* 31 (2006): 1177–1187.

7. F. Baldissera, P. Cavallari, L. Craighero, and L. Fadiga, "Modulation of Spinal Excitability During Observation of Hand Actions in Humans," *European Journal of Neuroscience* 13 (2001): 190–-194.

8. S. Strogatz, *Rhythms of Nature, Rhythms of Ourselves* (London: Allan Lane, 2003).

9. E. Tognoli, J. Lagarde, C. D. DeGuzman, and J. A. S. Kelso, "The Phi Complex As a Neuromarker of Human Social Coordination," *Proceedings of the National Academy of Sciences* 19 (2007): 8190–8195.

10. O. Oullier and J. A. S. Kelso, "Coordination from the Perspective of Social Coordination Dynamics,: *Encyclopedia of Complexity and System Science* (2009): 1–29.

11. V. K. Jirsa, and J. A. S. Kelso, *Coordination Dynamics: Issues and Trends* (New York: Springer, 2004).

12. A. Dijksterhuis and J. A. Bargh, "The Perception-Behavior Expressway: Automatic Effects of Social Perception on Social Behavior," *Advances In Experimental Social Psychology* 33 (2001): 1–40.

13. E. Hatfield, J. T. Cacioppo, and R. L. Rapson, *Emotional Contagion* (New York: Cambridge University Press, 1994).

You and I as one

Any social group can be thought of either as a collection of individuals or as a single new entity with emergent, unified group behavior. When a mob forms to surge together down a street one way and then another; when a flock of birds wheels about together, closely clustered as they fly without colliding; and when an orchestra performs with highly coordinated timing, we momentarily forget about the individuals and see the collective behavior as a new, single social entity. Indeed, as John Cacioppo discusses, many species seem to gather, flock, and coordinate to form such collectives. For humans, members congregate in this way in many situations, from flash mobs and sports teams to choirs and audiences.

The drive for people to affiliate and group is not sufficient on its own to produce the coordinated behavior that emerges from such a collective. Sometimes an organizing signal, like the conductor of an orchestra, can synchronize the behavior. Other times, common goals and behavioral constraints can synchronize a group, as in a flock of birds. In Chapter 6, "The suspension of individual consciousness and the dissolution of self and other boundaries," Gün Semin discusses how such synchrony may be self-organizing—that is, it is achieved without intention, effort, or awareness by our social brains, even when there is no clear signal or constraint. In cases of such human sociality, the group may act as though it has a single mind. Indeed, Semin approaches this issue to relate the collective behavior as an embodied consequence of individual social forces that jointly operate to satisfy our need to affiliate, and to consider how connecting behavior through synchrony may create a collective mind.

Howard Nusbaum specifically discusses a different invisible social force that has evolved with the power to bind people into a collective: language. Language is the richest social signal that has the power to move people to act, and to move groups to act together. For language to act as a force, it must somehow affect people with sufficient social and emotional impact. As Nusbaum discusses the impact of language, it operates at a social and emotional level similar to that discussed by Semin, rather than exclusively through the inferences drawn from meaning.

7*

Action at a distance:
the invisible force of language

Language is one of the most important ways in which the social brain makes connections, enhances connections, and severs connections among people. Language is our primary medium of social exchange, grounding and elaborating ourselves and our relationships in every conversation. However, language goes well beyond personal connections to

* The lead author is Howard Nusbaum, Ph.D., Professor of Psychology and Computational Neuroscience, and Codirector of the Center for Cognitive and Social Neuroscience at the University of Chicago. He has served as the Chair of the Psychology Department since 1997. He has served as the editor for the *International Journal of Speech Technology* and is on the editorial board of *Brain & Language,* and has edited several books on spoken language processing. His research interests include spoken language use, mechanisms of learning and attention, and the role of sleep in learning. His recent research has investigated

connect us culturally through stories, songs, and shared manners of speech. Language also provides the formal framework that defines many of our social institutions. Language gives form and substance to the governance and behavior of every social institution, from education to law, to religion. Clearly, language serves to knit us together in many ways, both formally and informally. For a linguist, all these uses can be analyzed in terms of the structure of sentences and their content. However, structure and content do not, on their own merits, provide a complete picture of how language can have the impact it does on our sense of social connection. How does language move us to act, change our feelings, and connect us to others? It seems unlikely that the impact of language is simply the result of dispassionate rational inferences and conclusions drawn from a logical analysis of sentences.

In 1976, Barbara Jordan, a Congresswoman from Texas, gave the commencement address at Brandeis University. Listening to her speak about the importance of public service and the importance of using talent and ability in service of one's country was an impressive experience. Her delivery was clear and not particularly dramatic, yet the force of her speech was riveting. It was sufficient to turn a graduating senior's mind from graduate school in psychology to (at least momentary) consideration of a career in government service. A student with the long-held intent of becoming a researcher and with no interest in politics, government, or public service might seem to be an immovable object. Yet in that moment, Jordan's speech had sufficient impact to make government service seem like the only path one would want to take or should ever consider.

the social use of language and the evolution of language. In addition, he has been working on neural mechanisms of reward and economic decisions.

We often think about language in terms of the information in newspapers, speeches, or reports. However, language is the basis of all our social relationships and institutions. We reward and praise with language, and we shun and punish with language, perhaps more often than with any other medium. In the recent election, Democratic candidates actually gave speeches outlining different views of the importance of language in our society. One candidate held that words are simply words and have the force that we give to them only by reasoning about them. The other candidate argued that speech has the power to move people to connect and act. Nusbaum was struck by this debate because it seems to him that the power of language goes well beyond what linguists and psychologists talk about as "meaning," and that understanding the meaning of language may depend on understanding the social and emotional impact of language. In this chapter, the impact of language at a distance is explored.

Although her points were argued well, the impact of Jordan's speech was not simply rhetorical. John F. Kennedy's "Ask not what your country can do for you..." and Martin Luther King, Jr.'s "I have a dream..." affected listeners deeply, well beyond the cognitive strengths of a good argument. Moreover, while all these speeches were delivered beautifully and from the heart, it is not the performance of these speeches that can move listeners to act on behalf of others. The performance alone cannot give substance to an empty message. Although there are cases in which a great performance may suggest briefly that there was content of import even in the absence of a real message, it is more likely the conjunction of message and delivery that moves people. In these speeches is a clear demonstration of the power of language. Language is more than words and delivery. Indeed, Cicero, in *De Oratore,* said that rhetoric conveys information, persuades listeners, and evokes emotion.

"In the beginning was the word"

If "the word made flesh" is taken metaphorically, the power of language can be made visceral in sermons. Consider the power of Jonathan Edwards' sermon, which Clark Gilpin discusses in Chapter 11, "Anthropomorphism: human connection to a universal society," to terrify a congregation and wrench them from complacency with images of torment. A sermon delivers a message, but it can do so in calm tones of instruction or with fire and brimstone. The choice and poetry of words and the cadence and intonation of speaking can draw the listener in slowly or seize the listener suddenly, the very sounds of speech painting images in the mind while igniting new inferences with literal and metaphoric descriptions.

In the realm of the spiritual, few corporeal manifestations can be perceived directly. Neither heaven nor hell, neither God nor the Devil can be seen, heard, or touched. Preaching is needed to spell out the work of unseen hands and will and to illuminate the power of the unseen. The force of that which is not seen can be felt only when transmitted directly through speech.

In *Phaedrus,* Plato described rhetoric as the art of leading the soul. Thus, it is not surprising that although the core of religion is a collection of beliefs and concepts and canons, the fabric and form of religion is language. Symbols and icons are certainly important, but language is the medium through which the force of theology is actualized in prayers, benedictions, sermons, and teaching. Language can reach across time and space to change minds, feelings, and behavior, encoding laws and beliefs and presenting them with a concrete reality in the here and now. This is one kind of impact from author to audience in which a kind of connection is constructed, bridging minds in the process.

In religious practices, another kind of connection is formed within a congregation. Joint recitation, responsive reading, collective listening, and understanding may serve to connect people in a religious service. Although joint participation in any event (such as sports or theater) may have some of the same effect as discussed by Gün Semin, the content of language and the intent of the messages in religious practice is often focused on developing and strengthening social and spiritual connections.

Everyone knows someone who moved to another country, or another part of their own country, where speech patterns differ. After a period of time, in the context of novel speech patterns, some people adopt the speech patterns around them. This kind of linguistic convergence is well documented and is moderated by social factors such as the desire to be accepted or the attempt to be persuasive.[1] When people talk together, one person's speech can impact the way another person talks to promote social connection. Indeed, the same kind of behavioral convergence is found over the course of conversations for other kinds of non-linguistic actions as well.[2]

Information, impact, and understanding

How does language bind us together and compel us to action, thought, and feeling? When we talk, sound vibration is transmitted from mouth to ears. Facial expressions and manual gestures punctuate, illustrate, and illuminate our speech. These acoustic and visual signals travel over space and time to the eyes and ears of the audience. The impact of such communication is true action at a distance.

This notion of language as action at a distance is relatively well accepted in the scientific study of language. But in research on language and communication by psycholinguists and linguists, the emphasis is on the information contained within an utterance, along with the structure and form by which this information is presented. Research questions often focus on the variety of ways the same message can be framed and how listeners interpret such messages. But little of this work addresses the impact of the message itself.

The standard view of language processing is that we hear the sounds of speech (acoustic patterns) that we translate mentally into words. The meanings of these words are determined and then the meanings are combined (through our knowledge of sentence structure) to result in sentence meaning. Given that a sentence typically occurs in a context of other sentences (such as a sermon) or in response to other speech (such as a conversation), this context is then used to frame and reinterpret the sentence meaning.

Metaphor, irony, sarcasm, and other figures are generally thought of as being understood later in this process, although in Chapter 11, the immediate power of these factors to affect us is quite clear. Some people believe that emotion and attitudes are not understood until after the linguistic message has been determined. The impact of language on attitudes through persuasion is viewed by cognitive psychology and linguistics as occurring after the message has been understood. However, this does raise the question about whether a "message" (to be understood) is simply the linguistic properties of a sentence or the impact of the linguistic form on an audience.

A very different view of language might come from considering vocal communications that often have a direct impact on an audience. Sometimes just listening to laughter compels listeners to laugh. Sometimes hearing a cry of terror directly imputes an instantaneous feeling of fear. Hearing someone weep can produce a feeling of sorrow and perhaps even cause one to cry. This suggests that some forms of vocal communication can elicit direct empathic sharing, as Jean Decety discusses in Chapter 9, "Empathy and interpersonal sensitivity." These forms of vocal communication can produce direct results in listeners without following a route of symbolic reference and interpretation.[3] Language has typically been viewed as operating by the more symbolic route because of the linguistic claim that symbols are not the things they stand for, and thus must be understood—words and sentences are not felt. However, the right insult, angry words delivered in the right way, or the right praise seems to be felt directly, perhaps through the same kind of mechanisms by which empathic sharing occurs.

But how is this achieved? This kind of impact of language is not a result of understanding the content of speech. It reflects social goals and motives. To the extent that this kind of social impact may parallel empathic processes, we might find that similar mechanisms are involved. Some kind of resonance must be established between a speaker and listener, and language can serve to establish this resonance and lead to subsequent action by the listener. This idea of a resonance in an audience does seem more compatible with the effects of vocal behavior, such as laughter or crying, on a listener than the symbolic interpretative view. This notion of resonance may also be useful in understanding other kinds of language impact, from the fiery sermon of Edwards to the passionate speeches of Obama, Kennedy, and King.

In listening to speech that has impact, language has created a state in the listener that reflects the intention of the speaker. Whether it is fear or connection, somehow language can operate as the medium by which social and emotional psychological states get transmitted to an audience. But how does this impact get created in the social brain?

Language impact in the social brain

Understanding spoken language has typically been viewed as an analytic process in which sound patterns are translated into linguistic symbols by the brain. However, an alternative theory is that we understand spoken language by using our motor system to simulate what might have been said. The idea is that trying to mentally produce the speech internally (without talking) might help us understand what is said when speech is not clear. However, this theory did not have much neural plausibility. People do not generally move their mouths overtly or even covertly while listening.

The discovery of mirror neurons, examined at greater length by Steve Small in Chapter 8, "Hidden forces in understanding others: mirror neurons and neurobiological underpinnings," suggested one kind of mechanism that might instantiate this kind of process. In certain parts of the brain that are involved in the control of action, some neurons that respond when making certain actions also respond when observing the same actions. This led to the inference that these neurons are involved in understanding the behavior of other people. The reasoning is simply that neural activity in the observer's motor system that results when seeing a behavior essentially establishes the brain state that would correspond if the observer were doing the same thing. Another way to say this is that there is a resonant response in an observer's motor system to observing a behavior, and this may potentiate a degree of social coordination and connection, as discussed in Chapter 6, "The suspension of individual consciousness and the dissolution of self and other boundaries." This kind of neural resonance has been shown to play a role in understanding speech when seeing a talker's mouth move, even if the listener does not actually make any mouth movements.

Mirror neurons demonstrate that observed action can produce a resonant response in an observer's brain. Seeing a talker's mouth move creates a motor-resonant response that aids in understanding speech sounds. However, these two observations are very different from the idea that the meaning of sentences can create a resonant response in the listener's brain. In the case of speech, mouth movements are the actions that create speech. This might seem like a very special case. The traditional linguistic view of language is that words and sentences are symbolic: Language describing action is not action itself. Language describing emotion is not the emotion itself. The entire concept of a symbol is that a symbol denotes something or stands for something, but the symbol is not the thing itself. But it now appears that this long-held notion may be wrong.

The idea that seeing an object or event gives rise to brain states that resonate with previous experiences of that thing suggests a mechanism for language understanding to go beyond symbolic interpretation. Given that there is a brain mechanism for reexperiencing actions or sensations, this same mechanism may operate even when there is just a symbolic linguistic description. Understanding language may take place by invoking such resonant past experiences in the brain.

For example, when listening to sentences about hockey action, hockey players show neural activity in their motor system, which is not seen for people who are naïve to hockey.[4] Experience playing hockey recruits the motor system in service of understanding hockey sentences as if one were watching or playing hockey when only listening to speech. This suggests one way in which language can have a direct impact on a listener. Rather than making inferences about actions based on the meanings of sentences, understanding a sentence may be a resonant motor system response in the listener to a description of an action. If this idea is extended more broadly, language impact may come from such resonant responses. Language understanding and the impact of language may result from processes more similar to the effects of hearing laughter.

Hearing a sentence may create in a listener a set of resonant responses very similar to the patterns that correspond to the actual situations being described.

Such resonant responses need not be confined to the motor system and actions. Emotions such as fear or joy may be empathically evoked in listeners by speech, just as a scream or laughter might. Verbal expression of attitudes may produce similar attitudinal responses in listeners. Moreover, if a listener's resonant response is strong, there may be increased empathic overlap with the speaker, which may serve to increase social connection. To the extent that people speak together and share feelings, social connection may increase as well.

Evolution of social connection by communication

The human social brain constructs connection and understanding by anthropomorphic projection, as discussed by Nick Epley in Chapter 10, "Seeing invisible minds." When we observe the behavior of nonhuman entities, anthropomorphic projection may form a feeling of social connection. Social connection depends in part on empathic responses to observed action. It has been argued that this same foundation is the basis for the evolution of language as well—observed action may be the foundation for language.[5]

A mirror neuron theory of the evolution of language starts with the assumption that we understand others by observing their actions. Communication depends on the regularization of typical actions that can be pantomimed. In principle, this could lead to a kind of manual gestural system of communication akin to sign language. Hand and arm movements can depict a wide range of actions, both by first-person depiction of an action (such as screwing the top onto a jar) and by third-person depictions (such as using the fingers to portray a person walking). Hand and arm postures can depict objects, such as a cupped hand representing a bowl. Combining sequences of such object and action depictions can communicate relatively complex messages even without a formal language. This is a far cry from sign languages in which the mapping between hand shapes and movements is not visually transparent in this way. But as the pressure to communicate a broader range of messages increases, a manual gestural language would have to be modified to reflect more abstract symbols, ultimately leading in the direction of a sign language.

However, this kind of manual language has one major drawback. Communication depends on visual contact. To understand a gestural message, it is necessary to see the hand movements. This limits the distances over which communication can take place. Moreover, a communication system that is effective for a group should not depend on face-to-face dyadic interaction, but should allow for more broadcast communications. There is a great survival advantage in being able to maintain social connection and convey information over distances that go well beyond face-to-face communication.

Many species of fish, amphibians, reptiles, birds, and mammals commonly exchange information at a distance through vocalizations. The learned songs (and some calls) of songbirds are particularly rich sources of information conveying, to the receiver, individual identity and a host of other characteristics of the sender. Some mammals exchange information at great distances through calling behavior. Humpback whale vocalizations are perceived over extremely long distances and may be used to maintain social groups at distances as great as 5km. African elephants can recognize friends and relatives from their calls at a distance of 2.5km. Human sheepherders keep each other company from the top of one mountain to another in the Canary Islands using a whistled language called Silbo Gomero. Whether for purposes of mating, threat, warning, or social organization; to convey information regarding location, identity, and motivation; or as directed at one individual or toward far-flung groups, vocal communication plays an important role in the social connection and behavior of a great number of vertebrate species. Thus, while human vocal communication is enhanced substantially in face-to-face interaction, the evolution of speech has resulted in a system that can function even in the absence of direct visual observation. The sound of a threat, warning, or distress can have substantial impact, even at a distance from the speaker.

Social impact and embodied language

Many animals use vocalizations to maintain social group structure and to provide information relevant to a group. Even very young children use the way people speak as a marker of their social affiliation.[6] Such vocalizations are typically not viewed as symbolic forms that must be decoded and interpreted. Instead, these vocalizations are mapped onto internal states more directly, much as human laughter may be. Although linguists and psychologists have often tried to differentiate human language from these kinds of vocalizations, language can impact us in much the same way. Perhaps language is less symbolic and more direct than scientists have thought.

What does it mean for language to have impact? If we take laughter as our model, perhaps it means that language gives rise to responses that have a direct effect on our perceptual and motor systems. Sermons may terrify because they create worlds inside of the listener that seem real. We can see and hear torment and imagine the feelings of pain and suffering. This is not a symbolic interpretation, but a real experience created from language. However, just as we can distinguish the pain we feel from the pain of others, even when we have strong empathic responses, we can distinguish such created experiences from those that occur in the real world. The stronger the language, the richer the imagery, the more intense the delivery, and the more salient the mentally created experience. One impact of language may be to create real feelings and sensations, and imagined movement and behavior.[7] When language hurts us through criticism, rejection, or insult, it may do so by activating our experiences of real pain. When we are soothed by language, it may produce the same kind of endorphin effect that placebo treatments can invoke. When language binds us together, it may do so by creating the kind of shared emotional states that characterize empathy.

In many respects, the linguistic view of language use does not engage these ways in which language functions. Linguistics treats language as consisting of patterns of symbols divorced from their origins in a human mouth, almost like print on a page rather than speech. But it is spoken language that the social brain evolved to use—print is a very modern invention which did not play any role in our evolution. In contrast, speech is produced from a coordinate action of muscles, compressing lungs and moving tongue and jaw under the control of neurophysiology and hormones. Unlike the shape of printed letters, the sound of speech is shaped by attitude, emotion, and intent, and its impact may be to transfer specific embodied states from the speaker to the listener.

Conclusion

Although physicists have debated the possibility of action at a distance for quite some time, the biological form of action at a distance is well established, achieved through vocal communication in an extremely broad range of behaviors and settings. The force of language is carried by the form, content, and delivery of a message. And the impact of this force may be created in the minds of an audience by the resonant invocation of past real experiences. Real pain and sorrow, real comfort and joy, and real love and caring are all part of our shared human experience. The impact of language may come by invoking resonant past experiences that can create a platonic mental moment that flickers with the shadows of those experiences. To understand how this process works, we need to study how brain mechanisms operate to translate the sounds of speech into the impact of language.

Endnotes

1. H. Giles, "Accent Mobility: A Model and Some Data," *Anthropological Linguistics* 15 (1973): 87–105.

2. J. Lakin, and T. L. Chartrand, "Using Nonconscious Behavioral Mimicry to Create Affiliation and Rapport," *Psychological Science* 14 (2003): 334–339.

3. F. Foroni, and G. R. Semin, "Language That Puts You in Touch with Your Bodily Feelings: The Multimodal Responsiveness of Affective Expressions," *Psychological Science* 20 (2009): 974–980.

4. S. L. Beilock, I. M. Lyons, A. Mattarella-Micke, H. C. Nusbaum, and S. L. Small, "Sports Experience Changes the Neural Processing of Action Language," *Proceedings of the National Academy of Sciences* 105 (2008): 13269–13272.

5. G. Rizzolatti and M. A. Arbib, "Language within Our Grasp," *Trends in Neurosciences* 21 (1999): 188–194.

6. K. D. Kinzler, K. Shutts, J. Dejesus, and E. S. Spelke, "Accent Trumps Race in Guiding Children's Social Preferences," *Social Cognition* 27 (2009): 623–634.

7. H. Shintel and H. C. Nusbaum, "The Sound of Motion in Spoken Language: Visual Information Conveyed by Acoustic Properties of Speech," *Cognition* 105 (2007): 681–690.

Systems and signals
for social coordination

How do we understand each other as people? Do we take people at their word, or do actions speak even louder? When moved to affiliate and to act in concert with other people as a group, we need to understand the communications and actions of others. Language can move us to action even across great distances, but how does it do so? Although we can observe the behavior of groups of people as coordinated, the mechanism of achieving this coordination is unseen. We may be driven socially to form groups, but how does that drive function in the individual to cause us to cohere? To go beyond the observation and experiences we have with groups and group behavior, we need to understand what makes the engine of social connection run. At one level, we can talk about language as a force itself, as Howard Nusbaum does. We can talk about the synchronization of individual behavior, as Gün Semin does. However, both of these are observations about the way individuals may become part of a group. To go beyond this, we must look to our biology to understand how the machinery underneath our sociality leads to connected minds.

Semin suggested one way our brains may seek to connect. Some neurons in an area of the brain that is involved in the control and planning of our actions also respond when we observe actions we have performed. Such neurons might be thought to "resonate" when seeing someone act or speak with our own experiences. Neurons that mirror actions and behavior have been thought to play a role in the process of understanding that behavior and the social connection that may form as a result of the resonance. In the next chapter, Steve Small discusses these neural mechanisms and how they may be important in helping us understand spoken language and, possibly, social behavior.

8 *

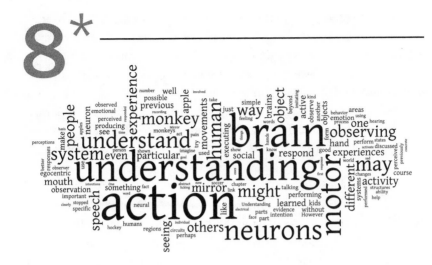

Hidden forces in understanding others: mirror neurons and neurobiological underpinnings

It is one thing to perceive objects in the environment and another to *understand* what is perceived. A rodent that senses an apple definitely has a notion that this represents something edible. A monkey might realize that the apple can be eaten but that it also can be thrown. A human might perceive it as a food, an object to be propelled, a temptation that

* The lead author is Steven L. Small, M.D., Ph.D., a Professor of Neurology and Psychology; Associate Chair for Research in Neurology; Member of the Committees on Neurobiology and Computational Neuroscience; and Senior Fellow, Computation Institute, at The University of Chicago. He is currently Director of the Human Neuroscience Laboratory and was founder of the Brain Research Imaging Center. He is an elected member of the American Neurological Association, a fellow of the American Academy of Neurology, and Editor-in-Chief of

should be resisted, or something that falls out of a tree at a specific acceleration. For each individual animal or person, understanding an apple means to take the sensory perceptions of the apple and to use previous experience and knowledge to fit it into an overall context. In this way, understanding a particular apple depends on our previously having seen, touched, and smelled apples; eaten them; read about them; and perhaps even been hit by a falling or thrown apple. All of our previous experiences come to bear every time we encounter a new perception that we must make sense of—and, of course, this represents virtually every moment of our waking lives.

Our perceptions vary enormously from seeing simple objects (such as apples), taking in more complex entities (such as restaurants or neighborhoods), hearing noises or speech, and seeing actions (such as simple manual actions or sporting events). A major question for brain research is, how can we possibly understand all these different kinds of input, and what brain circuits are used to do so? We assume that such an understanding means weighing these inputs against our previous experiences in some way and then integrating the new and old together. This integration requires constant and dynamic changes to the brain structures that represent what we know and how we use it.[1] One way this could happen is that when we perceive something new, we actually reenact in our mind's eye the previous related perceptions. For example, if we

the international journal *Brain and Language*. Small's research concerns the neural basis of human language and its breakdown after injury. He has published more than 120 scientific articles, primarily about human language, from the perspectives of artificial intelligence, cognitive psychology, computational neuroscience, human systems neuroscience, and clinical neurology.

Human language represents a unique product of our social species, and the tremendous evolution of the primate cerebral cortex simultaneously supported the development of both language and sociality. Language is the defining feature of our species. In his twelfth-century volume, *Guide to the Perplexed*, Maimonides viewed it as tautological that man is a speaking animal: "There is no third element besides life and speech in the definition of man." But how does the brain implement this unique function in the context of its common ontogeny with social function? This chapter discusses the possibility that the recently discovered "mirror neurons" of the cerebral cortex of macaque monkeys play a special role in the ability of humans to understand each other with language by using a mechanism of observation and covert emulation. If the neurobiology of language were partly grounded on such systems of visual observation and imitation, this would overlap integrally with the biology of the social brain.

encounter an apple—having encountered many previously—perhaps we actually imagine (seeing) one or more previous apples or episodes involving apples, or imagine (performing) one or more instances of biting an apple or throwing one, or imagine tasting and smelling one. We might also imagine hearing the word *apple,* producing the sounds of the word, seeing the written form of the word, or even hearing, seeing, or producing synonyms or related words in our first (such as "Granny Smith") or second (such as "pomme") languages. We propose that understanding an apple is tantamount to executing this entire set of processes and, thus, that the circuits for understanding are very complicated and take up a large portion of the brain.

Of course, not all of what we understand is directly available to the senses; we can clearly understand beliefs and emotions as well as physical objects and overt actions. Brain researchers generally take the view that previous experience guides the understanding of abstract concepts in much the same way that it guides the understanding of the more concrete entities. For example, we can understand the emotional states of other people by imagining being in those states ourselves. When I see someone feeling happy or sad, I can evoke examples from my own previous experience of feeling happy or sad (perhaps even for the same reasons), and by feeling the emotion, I can understand it. The closer my previous experience is to the perceived one, the better the "understanding." This principle holds whether one is trying to understand objects or people. An important distinction between people and objects when trying to understand their actions is that people, but not objects, have intentions.

The social brain

The critical question for neurobiologists is, how does the brain understand? In particular, we want to know whether the same brain circuits that are used when we experience things personally are also used when we try to understand another person having the same experiences— whether these experiences are concrete, such as grasping a cup or hitting a baseball, or more abstract, such as feeling sad or fearful, or in pain. We also want to know whether understanding the simple concrete perceptions and these highly complex emotional states are mediated similarly in the brain. Further, we are interested in the overlap between conscious and unconscious understanding and shared or personal understanding.

One way to address these questions is to examine the brain structures responsible for specific types of personal sensations, actions, and cognitive processes, and to see if these same ones play a role when individuals attempt to understand these functions in others. Using modern techniques of physiology, experiments of this type have been conducted in both monkeys and humans, with some surprising results.

It is now possible to measure brain activity of humans while having a wide range of different kinds of experiences. Such human neuroimaging experiments have suggested that understanding actions and objects invokes some of the same brain structures used to perform the actions and to act on the objects. With respect to actions in particular, humans sometimes use their own motor repertoire in interpreting actions, possibly by imagining or mentally simulating the perceived action. When people are asked to observe the actions of others, particularly goal-directed actions involving the hands or mouth, they seem to activate brain regions for moving the hands or mouth. Thus, there is a link between observing actions and executing actions. This has a relevance to education as well: "Understanding by doing" (by observing and then executing) has a long and valued tradition in American education,[2] and these recent scientific results might help us understand why this is effective.

When my son was five years old, he was a member of a kids' soccer team in Hyde Park, on the campus of the University of Chicago. His coach was a professor of history at the university, a woman who had never played soccer, but she was a voracious reader. And in her readings on the subject, she took careful note of all the methods needed to play soccer and the rules and regulations. She methodically took the kids through all the (theoretically relevant motor) steps needed to dribble the ball, to pass, and to shoot—flex your foot this way, bend your leg that way, keep your arms this way, and so on. The kids tried to follow the verbal instructions but their motor performance was less than stellar—they learned a little bit, but they lost all of their dozen games, for a depressing 0–12 record. The next season, the same team was coached by another volunteer parent, this time an engineer from Trinidad who had played soccer his whole life. The instructions he gave the kids were quite different: "Follow me and do what I do." There were no suggested foot flexions or extensions, and no specific leg movements proposed. The kids learned the skills and won all their games. Why? The kids learned by observing a good model and then imitating what they observed the person doing, which appears to be a way to learn motor skills that is far stronger than that of explicit motor instruction.

When people have strokes, a part of their brain dies and they can lose the ability to speak or use a hand properly. We are now using this idea of imitation in a treatment program to reeducate stroke victims to use their hands better and to pronounce words better. For these imitation-based treatments, people first observe a particular hand action or speech sample on a video monitor, and then they try to produce it. In fact, they observe it many times before they even try to do it. They are never told *how* to move their hands or mouths; they are just told to copy what they see. Over the course of six weeks, people with hand problems progress from imitating a simple grasp of a cup, to picking up a telephone and dialing a number, to picking up a toothbrush, brushing their teeth, and returning the brush to the sink. Those with speech problems progress from imitating simple words of a single syllable to longer, less common words and even short phrases. We have already shown that these therapies have beneficial effects in a number of people and are now trying it out more extensively.

There seems to be an important link between observing and executing actions. Similarly, there is a link between observing action and understanding emotion. This link has been most clearly demonstrated in the case of facial expressions—if I observe the muscles of the face in a position to convey an emotional state, clearly I perceive the emotion. What has been shown recently is that observing such facial expressions leads to two kinds of brain activations in the observer: The first set of regions activated are those that would be used by the observer to execute the identical face movements, just as with hand or mouth movements. However, additional regions are also active, and these are precisely the ones that would be involved if the observer were to feel the observed emotion personally. Thus, the circuitry for action observation in the human brain is interdependent with parts of the brain critical for understanding more complex nuanced aspects of the world.

Mirror neurons

It turns out that there may be cellular building blocks in the brain that are particularly important for observing and executing actions, and may ultimately lead to an explanation of action understanding and imitation-based learning. In fact, such structures would contribute to any form of understanding that could be partly explained by imagined reenactment of perceived actions (such as seeing an emotional facial expression or hearing a cry of pain). The cells under discussion are a type of nerve cell, or neuron, discovered in the front part of the monkey brain by Professor Giacomo Rizzolatti and his colleagues at the University of Parma. The scientists trained monkeys to perform specific actions such as grasping an object or licking their lips, and conducted electrical recordings in regions in the front of the brain known to coordinate movements. These recording machines note brain activity both visually, as a graph on a screen, and auditorily, by a loud series of clicks, indicating the firing of a neuron. Rizzolatti and his team were focusing on a particular region in the front of the brain and were having the monkey perform all sorts of hand, mouth, and eye movements to see how the brain cells were organized to make these movements happen. One day (or so the story goes), one of the researchers returned from lunch while the electrical recordings were being made. He was finishing a cone of superb Italian gelato when the recording device suddenly starting making a loud series of clicks. The returning scientist stopped licking his ice cream cone to see what was going on and the noise stopped. When he restarted licking his gelato, the clicks resumed; when he stopped again, they stopped. The investigators had discovered a type of neuron that was sensitive to the monkey observing a particular human action.

It was not surprising that after training to perform an action, some neurons in the motor region of the brain responded while performing that action, whereas the same neurons would not have responded beforehand. However, it was extremely surprising to find that some of those neurons also responded vigorously when the monkey *observed* the very same learned actions. Through a methodical and systematic approach, this group was able to make a more elaborate and far-reaching set of observations. For a small subset of neurons, if a monkey had learned to reach for a particular object, seeing another monkey reach for the same object would cause the neuron to fire. For a different subset of neurons, if the monkey had learned to pucker his lips, the neuron would fire when the monkey did this or when the monkey observed another monkey (or even a human) doing this. These motor neurons have been dubbed "mirror neurons" because they respond during both execution of action and during observation of the same action in a mirrorlike fashion.[3] Mirror neurons are not active during observation of an appropriate action if there is no goal (when the object is absent) or when an appropriate object is presented alone. Mirror neurons have been discovered both for mouth actions and hand actions, and for both visual and auditory perception of actions.

Seeing a previously learned action performed by someone else seems to resonate in some neurons in the motor system almost as if the action were being performed by the observer. It is as if the observed action stimulates some motor neurons to "remember" what it was like to perform that action. Of course, this is not memory in the overt sense of conscious recollection, but rather that the experience of execution changes the response of the neurons to observation. Not all the motor neurons respond this way, but a small number have been shown to respond when performing and observing an action. Such mirror neurons could provide a correspondence between the experience one has of performing an action and seeing the same action performed by others.

These mirror neurons might provide one basis for understanding action. Relating actions we observe to actions we have carried out seems like an important component for comprehension. After all, we knew what we were doing when we performed an action. If that experience is somehow reinstated during observation, we might attribute our past experience as the interpretation of the present observation. Imitation when observing an action might occur because our motor system is stimulated by observing an action. Coordination of action could occur because, in representing others' actions as if they were our own, our brains may be able to compute the time when we can act without disrupting the other person. This is the kind of process that is described in Chapter 6, "The suspension of individual consciousness and the dissolution of self and other boundaries," when Gün Semin describes how groups of people can synchronize their actions such as clapping together.

From monkey brains to human intention

Of course, relating responses in monkey brains to human brains is neither direct nor simple. Parts of the monkey brain and parts of the human brain that putatively correspond, although similar, are not identical in number, size, or location, and probably do not do exactly the same things, because monkeys and humans have evolved to have somewhat different capacities and behaviors. Furthermore, the study of mirror neurons in monkeys is based on recording the responses of individual neurons, which is not generally possible in humans except in rare cases of medical necessity. The measures we can make on intact human brains come from the responses of many thousands of neurons, so it is difficult to make claims about neurons that respond in producing an action and perceiving the same action. This means that any claims about human mirror neurons depend on a degree of good faith and inference rather than specific empirical demonstration that individual neurons respond to both observing and executing action.

Researchers have measured human brain responses using a variety of methods, such as functional magnetic resonance imaging (fMRI), which demonstrate reliable effects of changes in neural activity by changes in blood flow. Although slower to respond than measures of electrical activity, fMRI provides evidence about where neural activity in the human brain occurs. This kind of research does show that observing action produces activity in areas of the human brain more typically associated with executing action. Although there may be some disagreement about which motor areas of the human brain are active while observing or imitating action and how these would correspond to areas in the monkey brain, there is good agreement that the human motor system responds for observation and imitation of action.[4]

There is quite a difference between recognizing an action and understanding that action. We can see a hand move through space with an open palm oriented with the flat of the palm moving toward the surface of an object, and predict where the hand will strike the object and that it will apply force to the surface of that object. But understanding the same general action as pushing a door open with the intent to enter versus slapping a person in the face is quite different. A ball can be thrown in a game of catch or can be a missile intended to do harm. The actions may be similar, but the intentions are different. Therefore, it is important that some researchers have argued that mirror neurons respond to the intention as well as the action.[5] However, to date, there has been no clear evidence that such neurons respond to intention—just that the sight of the action of reaching without an object to grasp, the sight of the object alone, and the sight of the action with the object present show different patterns of brain response. The fact that such different visual experiences lead to different patterns of brain activity does not provide clear evidence that intentions or goals are somehow part of the mirror neuron response to observed action.

Understanding spoken language as action understanding

The potential ambiguity of action is perhaps clearest if we consider language. Talking is a form of action, and understanding speech might be a form of action understanding. In talking, mouth movements are made in such a way as to create sounds that will have some effect on the listener, as discussed in Howard Nusbaum's Chapter 7, "Action at a distance: the invisible force of language." Listeners must understand what was meant by making those sounds. However, if someone says, "It's hot in here" or "You are a great friend," there can be ambiguity about the meaning. Such sentences could be straightforward observations, as they seem to be, or they could be something very different. We can ask to have a window opened or we can make a negative social comment using exactly the same sentences.

Nonetheless, there is good reason to believe that the human action system is involved in understanding speech and producing it. Although some ambiguities in speech or behavior simply cannot be resolved without broader contextual knowledge, the motor system may be important in understanding. If you try to have a conversation in a noisy bar, looking at your friend talking makes it easier to understand what is being said. We have shown that the motor system contributes to this process of recognizing speech. When listeners can see someone's face while talking, mouth movements increase activity within the motor system measured using fMRI. Furthermore, it is possible to show that the same parts of the motor system can be active in talking and in understanding speech. By analyzing which parts of the brain are active and when they become active, it is possible to show that when motor system activity in the frontal lobe of the brain precedes activity in other regions (such as the temporal lobe), listeners have a better understanding.[6] It is as if the motor system "recognizes" the speech before other parts of the brain when the talking mouth is visible to the listener.

This study shows that more information about the action of producing speech (visible mouth movements) can activate the motor areas of the brain during understanding of speech. But it is also the case that understanding speech without seeing the speaker depends on the motor system. Hockey players are experts at hitting slap shots, blocking passes, and whacking each other in the head with hockey sticks. They have done these things in the real world just as monkeys in Parma have learned to reach for certain objects. When hockey players listen to sentences describing hockey action, even without seeing the speaker or the action, motor areas are active during sentence understanding and these same motor areas are not active in people without hockey experience.[7] Understanding described actions appears to be influenced by the motor systems of people who have experience with those actions.

Reading minds through action

In understanding other people, we start with what we understand about ourselves. As Nick Epley describes in Chapter 10, "Seeing invisible minds," we take this kind of egocentric perspective in understanding other people, pets, God, or the behavior of inanimate objects. We know what we meant when we say something or do something, and we make the same attribution to others, even nonhuman others. Although this may be a good starting point for religions to help people feel connected to God, Clark Gilpin states in Chapter 11, "Anthropomorphism: human connection to a universal society," that it may not be uniformly informative about human behavior. Behavior is not transparent for the intention of that behavior. The fact that any particular action is not necessarily unique to the intent behind it is the basis of a great deal of misunderstanding in daily interactions. As a result, even if mirror neurons help our brains recognize actions and sometimes interpret them, there are real limits to how experience-producing actions can correctly inform social understanding.

In spite of this limitation, we may often do just this—assume we understand another's actions because of what we would intend, were we to do the same thing in the same circumstance. Human social understanding does suffer egocentric limitations often, and to the extent that it does, something such as a mirror neuron system may play a role. For example, as discussed by Jean Decety in Chapter 9, "Empathy and interpersonal sensitivity," our ability to understand the pain of others may derive, in part, from the neural systems involved in our experience of pain, but goes beyond this starting point. The social brain is, on one level, perhaps, a very egocentric brain. But the fundamental motivation to connect with others has resulted in systems built on top of these egocentric foundations. If social understanding depended entirely on past experiences of our own intentions and actions, there might be much more misunderstanding and cynicism in the world. However, the capacity to reason, hypothesize, and model possible futures may increase social understanding beyond the anchor of purely egocentric perspective. We can conceive of alternatives to our own goals and motives and relate those alternatives to the actions we observe. To some extent, this process might also involve the motor system by mentally simulating actions and anticipated responses. By imagining how we might act in some situation to achieve a goal or the alternative ways we may act, given some intention, it may be possible to go beyond the limits of our own experience. Such constructive imagery may well depend on the motor system, along with other neural systems, but currently there is no scientific evidence that such a system might be linked with the operation of mirror neurons.

Conclusion

We are equipped to understand the world around us by relating what we perceive to our own experiences. With respect to actions in particular, our brains have specialized circuitry to relate previously executed actions to newly perceived ones, possibly by performing an internal (imagined) simulation of them. There is evidence, too, that we might understand the emotional states of others by a similar kind of process, whereby our brains activate circuits for experiencing the emotion as a way to understand that emotion in others. These brain mechanisms might also apply (to some degree) when we try to understand actions or feelings by non-human animals or even inanimate entities. This could be a partial biological explanation of anthropomorphism, as discussed by Epley and Gilpin. Of course, as humans we have the ability to go beyond these strict egocentric limitations and recognize and respond to our social connections more explicitly. This ability to go beyond the more basic grounding of the way we understand others may subserve part of the goal of some religions in fostering a more abstract view of our connection to others (see Kathryn Tanner's Chapter 13, "Theological perspectives on God as an invisible force"). Although a mirror neuron system might help form the basis for some aspects of social understanding, there may be other invisible forces at work supported by these and other neural systems in our social brain.

Endnotes

1. U. Hasson, H. C. Nusbaum, and S. L. Small, "Task-Dependent Organization of Brain Regions Active During Rest," *Proceedings of the National Academy of Sciences of the United States of America* 106, no. 26 (2009): 10841–10846.

2. J. Dewey, "Democracy in Education," *The Elementary School Teacher* 4, no. 4 (1903): 193–204.

3. V. Gallese, L. Fadiga, L. Fogassi, and G. Rizzolatti, "Action Recognition in the Premotor Cortex," *Brain* 119, no. 2 (1996): 593–609.

4. M. Iacoboni, R. P. Woods, M. Brass, H. Bekkering, J. C. Mazziotta, and G. Rizzolatti, "Cortical Mechanisms of Human Imitation," *Science* 286, no. 5449 (1999): 2526–2528.

5. M. Iacoboni, I. Molnar-Szakacs, V. Gallese, G. Buccino, J. C. Mazziotta, and G. Rizzolatti, "Grasping the Intentions of Others with One's Own Mirror Neuron System," *Public Library of Science Biology* 3 (2005): 529–535.

6. J. I. Skipper, V. van Wassenhove, H. C. Nusbaum, and S. L. Small, "Hearing Lips and Seeing Voices: How Cortical Areas Supporting Speech Production Mediate Audiovisual Speech Perception," *Cerebral Cortex* 17 (2007): 2387–2399.

7. S. L. Beilock, I. M. Lyons, A. Mattarella-Micke, H. C. Nusbaum, and S. L. Small, "Sports Experience Changes the Neural Processing of Action Language," *Proceedings of the National Academy of Sciences* 105 (2008): 13269–13272.

Connecting and binding social brains and minds

We evolved as social organisms in the context of face-to-face interaction. Our human biology was established before the technology of email and cellphones. As a result, our biological nature is tuned primarily to social signals and interaction that occurs in the presence of another. It takes just a moment of observation for us to know a lot about another person as a social being. We can understand other speakers easily within tens of milliseconds of experience. Our brains have developed to make this kind of social connection quickly and easily, whether through language or action. To understand someone else, we need to be able to understand their goals and intentions. Moreover, we need to relate their behavior to our own individual and personal experience. Steve Small discusses some of the brain machinery that may allow us to translate our perception of language and observation of behavior into a form that can be related to our own use of language and action. This kind of resonance with experience, rather than elaborate inferences, may allow us to connect quickly with others, satisfying our drive for sociality.

But understanding behavior and communication is not all there is to forming social connections. It is one thing to read intentions and another to feel someone's pain. If we could understand only action and communication, an important element of human connection would be missing. In forming a collective mind, as Gün Semin discusses it, we need to have collective emotional responses. Jean Decety discusses the foundations of empathy and the brain machinery that supports this important capacity by providing a resonant response to the observation of another person's distress. Although sympathy may motivate helping others, empathy may be one of the social glues that binds us together as a collective social organism.

9*

Empathy and interpersonal sensitivity

A young girl, after watching a televised documentary account of child hunger and suffering in Bangladesh, pleads with her parents to "do something" to help them. A new child in daycare is being ignored by the other children with the exception of one little boy, who takes her hand and includes her in all his activities. These modest beginnings signal the important and ongoing role of empathy for surviving and flourishing in

* The lead author is Jean Decety, Ph.D., the Irving B. Harris Professor in the Departments of Psychology and Psychiatry, and Codirector of the Brain Research Imaging Center at the University of Chicago Medical Center. He is the editor of the journal *Social Neuroscience*. His interests include the investigation of the neurobiological mechanisms underpinning interpersonal sensitivity, particularly empathy and sympathy. His recent work focuses on developmental neuroscience with both typically developing children and adolescents, and children with deficits in empathic responding, such as antisocial behavior problems. Decety has published more than 115 scientific papers and recently edited the

our social world. It has been suggested that empathy is essential for navigating the social world; it enhances our understanding of others, improves the effectiveness of our social communication, and fosters mutually satisfying social relationships. Conversely, lack of empathy, or its maladaptive use, causes social relationships to falter and fail.

Empathy refers to our natural capacity to quickly and automatically relate to the emotional state of another person. Rudimentary forms of empathy appear early in the life course, and, with maturation, empathy can be experienced by simply reading about or imagining someone else's emotion. Empathy comes so naturally that physicians must learn to dampen their empathic pain responses when inserting a needle into a patient.[1] Just because empathy comes naturally does not mean, however, that it is instantiated in a discrete brain module that is automatically activated when one witnesses another's distress or suffering. Rather, the experience of empathy is underpinned by the combined activity of several dissociable psychobiological systems. Furthermore, empathy can be modulated by various contextual, dispositional, and interpersonal factors. The study of empathy—using sophisticated methods from psychology, neurology, neuroimaging, and neuropsychology—provides a unique opportunity to understand the invisible power of empathy in shaping our obligatorily gregarious social nature.

Defining empathy and its functions

Empathy can be defined as our natural capacity to share, appreciate, and respond to the affective states of others. This capacity is essential for the regulation of social interactions. For instance, empathy is believed to motivate prosocial behavior and inhibit aggressive behavior.[2] In addition, our ability to share the emotions of those we observe binds us to each

Social Neuroscience of Empathy (MIT Press, 2009) and *Interpersonal Sensitivity: Entering Others' World* (Psychology Press, 2007).

Empathy and sympathy play crucial roles in much of human social interaction and are necessary components for healthy coexistence. Sympathy is thought to have a key role in motivating prosocial behavior, guides our preferences and behavioral responses, and provides the affective and motivational base for moral development. Although the study of these abilities has traditionally been examined using behavioral and clinical methods, recent work in social neuroscience has begun to provide compelling and novel insights on the neural mechanisms involved in interpersonal sensitivity. These developments are explored in this chapter.

other and fosters a collective social identity. Empathy's invisible power is that it moves us to cooperate, coordinate our behaviors, and provide the needed care for one another. Notably, however, empathic concern does not necessarily lead to empathic behavior. First, empathy poses a paradox because sharing of feelings does not necessarily imply that one will act, or even feel impelled to act, in a supportive or sympathetic way. Second, the complexity of the social and emotional situations eliciting empathic concern influences the probability and nature of the help provided. Whether and how empathic actions are expressed depends on the feelings we perceive in the other person, our relationship with that individual, and the context in which we share an emotional state.

Empathy is critical for complex human interactions, but this does not mean that empathy and prosocial behavior have suddenly appeared with *Homo sapiens*. If empathy is a potent invisible force generated by the social brain, some form of emotion sharing should also be evident in other social species, such as nonhuman primates. Indeed, field observations conducted by comparative psychologists and ethologists suggest that behaviors homologous to empathy can be found in nonhuman primates.[2] Some have argued that empathy is not an all-or-nothing phenomenon, and that many intermediate forms of empathy exist between the extremes of mere agitation at the distress of another and full understanding of their predicaments.[3] Many comparative psychologists view empathy as a kind of induction process by which emotions—both positive and negative—are shared, and which increase the probability that the protagonists will subsequently engage in similar behavior.

Although certain nonhuman primates may share feelings between individuals, humans seem to have the unique ability to intentionally "feel for" and act on behalf of other individuals whose experiences may differ greatly from their own. Such a capacity may help explain why empathic concern is often associated with prosocial behaviors, such as helping kin, and why it has been considered the foundation for altruism—the expression of empathy and caring for those who are not kin. Evolutionary biologists have suggested that empathic helping behavior evolved because of its contribution to genetic fitness (kin selection). In humans and other mammals, an impulse to care for offspring is almost certainly genetically hard-wired. Less clear, however, is whether an impulse to care for siblings, more remote kin, and similar nonkin is genetically hard-wired. The emergence of altruism is not easily explained within the framework of

neo-Darwinian theories of natural selection (but see Cacioppo's information about this topic in Chapter 2, "The social nature of humankind").
Social learning explanations of kinship patterns in human helping behavior are thus highly plausible. Indeed, one of the most striking aspects of
human empathy is that it can be felt for virtually any "target," even targets of a different species (animals included). We can see a deer hurt by
a passing car or dogs locked in crates at a shelter and feel strongly for
their pain or confinement and future. In part, this kind of empathic
extension may be motivated by the kind of anthropomorphic attitudes
we have about nonhuman entities, as discussed by Nick Epley in Chapter 10, "Seeing invisible minds." The fact that we are adept at "feeling
for" others who are very different and who we can observe but not truly
understand suggests that we possess a capacity to cognitively rerepresent
others in our mind in a way that we can understand. Indeed, second-
order representation is a key component of empathy in humans and may
be a useful adaptation for human survival because it maximizes the range
of individuals with whom we can form a social bond.

The components of empathy

The psychological components that make up full-blown empathy are
supported by distinct and separable psychobiological systems. Empathy
can be decomposed into an affective component that includes the perception and sharing of an emotional state observed in another individual,
and a cognitive component that includes the motivation and intention to
respond. Closely related is a regulatory component that involves adjustment of one's emotional and behavioral response. The affective, cognitive, and regulatory aspects of empathy involve interacting yet partially
non-overlapping neural circuits. The initial component in the overall
process leading to empathy draws on somatic mimicry, also known as
"emotion contagion." This affective component of empathy develops
earlier than the cognitive component. Affective responsiveness is present
at an early age, is involuntary, and relies on mimicry and linking of
actions perceived in others with actions in oneself (perception-action
coupling). For instance, newborns and infants become vigorously dis-
tressed shortly after another infant begins to cry. Facial mimicry of basic
emotional expressions also contributes to affective sharing, and this

phenomenon starts very early in life, by approximately ten weeks of age. This primitive mimicry mechanism may be based on mirror neurons, which are sensorimotor neurons found in the premotor, motor, and posterior parietal cortex of the brain that become active when observing and when enacting a behavior, as discussed by Steven Small in Chapter 8, "Hidden forces in understanding others: mirror neurons and neurobiological underpinnings." This kind of mechanism may contribute to the development of empathy in the early preverbal period and continues to operate past childhood. Evidence shows that when we perceive emotions and actions of others, we use the same neural circuits as when we produce the same emotions and actions ourselves (for example, watching another individual being disgusted and experiencing disgust in oneself activate similar neural circuits). For instance, viewing facial expressions triggers expressions on one's own face, even without explicit identification of what we're seeing.[4]

The cognitive component of empathy is closely related to processes involved in "theory of mind" (the ability to attribute mental states to others and to understand that others' mental states can differ from one's own) and self-regulation. The capacity for two people to resonate with each other emotionally, prior to any cognitive understanding, is the basis for developing shared emotional meanings, but it is not enough for mature empathic understanding and concern. Such an understanding requires the observer to form an explicit representation of the feelings of another person, a process that involves additional mechanisms beyond the sharing of emotion and includes self-regulatory mechanisms to modulate the observer's experience of negative arousal. Specifically, to understand the emotions and feelings of others in relation to oneself, second-order representations of the other must be consciously available and must not confuse the other with the self. The medial and ventromedial prefrontal cortices are known to play crucial roles in decoupling first-person and third-person information and in maintaining representations of the other as distinct from the self.[5]

The regulatory component of empathy, especially regulation of internal emotional states and processes, is particularly relevant to the modulation of vicarious emotion and the experience of empathy as well as sympathy. Empathy is unlikely to lead to helping behavior if the observer is incapacitated by strong empathically evoked emotions, which is why emotional regulation is an important component in empathy. Indeed, children high in effortful control show greater empathic concern, and the tendency to experience empathy and sympathy versus personal distress varies as a function of their ability to regulate their emotions more generally.

How we perceive other people in pain

When witnessing another person experiencing pain, the scope of an observer's reaction can range from concern for personal safety (including feelings of alarm, fear, and avoidance) to concern for the other person (including compassion, sympathy, and caregiving). The existence of the perception-action coupling mechanism apparent in emotional contagion also seems to account for our ability to perceive and understand the pain of others. In the case of pain, individuals are predisposed to find distress of others aversive and learn to avoid actions associated with this distress. This is even the case in many mammalian species, including rodents. For instance, rats that had learned to press a lever to obtain food would stop doing so if their response was paired with the delivery of an electric shock to a visible neighboring rat.[6]

Recently, a handful of functional neuroimaging studies performed with healthy human volunteers revealed that the same neural circuits implicated in processing the affective and motivational aspects of pain in oneself account for the perception of pain in others.[7] In one study, participants in a magnetic resonance imaging (MRI) scanner either received a painful stimulus or, in other trials, observed a signal that their partner, who was present in the same room, would receive the same stimulus. First-hand experience of pain resulted in activation of the somatosensory cortex, which encodes the way we feel aspects of a noxious stimulus, such as its bodily location and intensity. Furthermore, the anterior medial cingulate cortex (ACC) and the anterior insula were activated during both first-hand pain and the anticipated experience of pain in someone else. These regions are responsible for the affective and motivational processing of noxious stimuli, such as those aspects of pain that pertain to desires, urges, or impulses to avoid or terminate a painful experience. A number of other neuroimaging studies of empathy for pain in adults and in children have demonstrated that the somatosensory cortex is activated not only during first-hand pain, but also during the perception of other people in pain. Altogether, strong evidence suggests that perceiving the pain of others triggers an automatic somatic sensory–motor mirroring mechanism between other and self, which activates almost the entire neural pain matrix, including the periaqueductal gray (a major site in pain transmission and processing of fear and anxiety) and the supplementary motor area that programs defensive movements in response to anticipated pain. Such a neural resonance mechanism provides a functional bridge between first-person and third-person information. It is grounded in the equivalence of self and other, which allows for analogical reasoning, and offers a possible, yet partial, route to understanding others.

Of course, human empathic abilities are more sophisticated than simply yoking perceptions of the self and other. In the eighteenth century, Scottish philosopher and economist Adam Smith proposed that, through imagination, "We place ourselves in his situation...enter as it were into his body, and become in some measure the same person with."[8] By means of imagination, we come to experience sensations that are generally similar to, although typically weaker than, those of the other person. This capacity to engage in role-taking has been theoretically linked to the development of empathy, moral reasoning, and, more generally, prosocial behavior. Unlike the motor mimicry and emotional contagion aspect of empathy, perspective taking develops later, possibly because it draws heavily on the maturation of executive functions (processes that serve to monitor and control thought and actions, including self-regulation, planning, cognitive flexibility, response inhibition, and resistance to interference), functions that are predominantly centered in the prefrontal cortex that continues to mature from birth to adolescence. Theoretically, imagining the other is distinct from imagining the self: The former may evoke empathic concern (defined as an other-oriented response congruent with the perceived distress of the person in need), while the latter induces both empathic concern and personal distress (a self-oriented aversive emotional response, such as anxiety or discomfort). This distinction has been supported empirically. When individuals are asked to imagine how they would feel in reaction to emotion-laden familiar situations and to imagine how a known person would feel if she was experiencing the same situations, common neural circuits are activated for both the self and the other. However, relative to imagining the self, imagining the other results in specific activation of parts of the frontal cortex that are implicated in executive control—the use of attention, working memory, and decisions—and an area of the brain at the interface of the temporal and parietal lobes of the brain that is a key component of a larger network of neural circuits involved in attention (sometimes called the temporoparietal junction).

Some researchers have hypothesized that the role of the frontal lobes and the temporoparietal junction is to hold separate perspectives or to resist interference against attention to one's own perspective. In a recent functional brain imaging study, participants were shown pictures of people with their hands or feet in painful or nonpainful situations and were instructed to imagine themselves or to imagine another individual experiencing these situations.[9] During perception of painful situations, both the self-perspective and the other perspective were associated with activation in the neural network involved in pain processing. These results reveal the similarities in neural networks representing first- and third-person information. However, the self-perspective yielded higher pain ratings and involved more extensive activation of some circuits in the pain matrix than did the other perspective, thus highlighting important differences between self and other perspectives.

Neuroanatomical regions and circuits form the foundation for the experience of pain in others, but they are not sufficient to explain variability in interpersonal sensitivity. Although empathetic brain circuits are activated by the mere perception of pain in others, activity in these circuits can be modulated by social, motivational, and cognitive factors. For example, observing pain in likable others (for example, those who played a game fairly) resulted in an enhancement of empathic brain responses, whereas pain in dislikable others (who played unfairly) did not. Another functional MRI study found that participants showed significantly greater responses in neural regions that are involved in pain perception when observing the pain of people who were not responsible for their stigmatized condition (for example, individuals who contracted AIDS as the result of a blood transfusion) than either controls (healthy individuals) or people who were held responsible for their condition (those who contracted AIDS through illegal drug use).[10] In addition, participants expressed more empathy and personal distress in response to the pain of people who were not responsible for their stigmatized condition as compared to controls. The level of empathic response, therefore, seems to be influenced by motivational factors and the interpersonal relationship between the target and the observer.

Altogether, these findings demonstrate that the similarities between affective representations of the self and the other stem from shared neural circuits that can be emulated either automatically or intentionally by the act of perspective taking. Importantly, these findings also point to some distinctions between these two representations, distinctions that contribute to our capacity to detach ourselves from others sufficiently to make considered responses to their pain.

Conclusion

Empathy, the natural capacity to share, appreciate, and respond to the affective states of others, plays a crucial role in much of human social interaction from birth to the end of life. As would be expected if empathy functions to enhance social cohesion, social nonhuman primate species also exhibit rudimentary versions of empathy. Humans abundantly have higher-level cognitive and social abilities (language, theory of mind, and executive functions) that can be deployed to modulate empathic responses, and that are amenable to modulation by lower-level processes such as emotional contagion and mimicry. These levels of processing enable empathy to have an impact on a wide variety of human behaviors. From motivating prosocial behavior to providing the affective and motivational bases for moral development, empathy is an invisible force to reckon with when considering how humans behave toward each other.

Endnotes

1. Y. Cheng, C. Lin, H. L. Liu, Y, Hsu, K. Lim, D. Hung, and J. Decety, "Expertise Modulates the Perception of Pain in Others," *Current Biology* 17 (2007): 1708–1713.

2. F. De Waal, *The Age of Empathy: Nature's Lessons for a Kinder Society* (New York: Harmony Books, 2009).

3. F. B. M. De Waal, "Darwin's Last Laugh," *Nature* 460 (2009): 175.

4. A. N. Meltzoff and J. Decety, "What Imitation Tells Us about Social Cognition: A Rapprochement between Developmental Psychology and Cognitive Neuroscience," *The Philosophical Transactions of the Royal Society* 358 (2003): 491–500.

5. J. Decety and J. A. Sommerville, "Shared Representations between Self and Others: A Social Cognitive Neuroscience View," *Trends in Cognitive Sciences* 7 (2003): 527–533.

6. R. M. Church, "Emotional Reactions of Rats to the Pain of Others," *Journal of Comparative and Physiological Psychology* 52 (1959): 132–134.

7. J. Decety and C. Lamm, "Empathy and Intersubjectivity," in G. G. Berntson and J. T. Cacioppo (eds.) *Handbook of Neuroscience for the Behavioral Sciences* (Hoboken, N.J.: Wiley, 2009).

8. Adam Smith, *The Theory of Moral Semtiments,* Sálvio M. Soares (ed.), MetaLibri, 2005, v1.0p.

9. P. L. Jackson, E. Brunet, A. N. Meltzoff, and J. Decety, "Empathy Examined through the Neural Mechanisms Involved in Imagining How I Feel versus How You Feel Pain: An Event-Related fMRI Study," *Neuropsychologia* 44 (2006): 752–761.

10. J. Decety, S. C. Echols, and J. Correll, "The Blame Game: The Effect of Responsibility and Social Stigma on Empathy for Pain," *Journal of Cognitive Neuroscience* (2009).

Seeing into my mind and other minds

Empathy is defined by Jean Decety as "the natural capacity to share, appreciate, and respond to the affective states of others." Empathy rests on our ability to see into the mind of another while distinguishing it from one's own to be in a position to cooperate, coordinate, and provide the needed care for others. The possession of empathic capacity is not sufficient to determine the precise nature of the response toward others, however. As Decety points out, whether an individual attends to and responds empathically upon observing emotion in another individual depends on, among other things, dispositional tendencies, the relationship between the individuals, and contextual constraints. Motivation to help another is also influenced by the amount of cognitive effort we are willing and able to exert to take the perspective of the other.

Perspective taking is essentially an attempt to see into the invisible mind of another. What we can't see, we model based on our own mind and likeminded individuals. Nick Epley shows that seeing into and connecting with other minds is such a frequent operation of the social brain that the absence of others inclines people to see human minds in nonhuman entities. Whether it's a tree, a pet, or God, ascribing mind to others endows them with the capacity to experience the same affective states we experience. This shared capacity evokes in us a tendency to feel and express empathic concern for their well-being. Unfortunately, humans do not grant all individuals, much less nonhuman entities, an equivalent degree of mind. Epley articulates how differences in the capacity to see mind in others have consequences for the way we feel about, think about, and behave toward others.

10 *———————————————

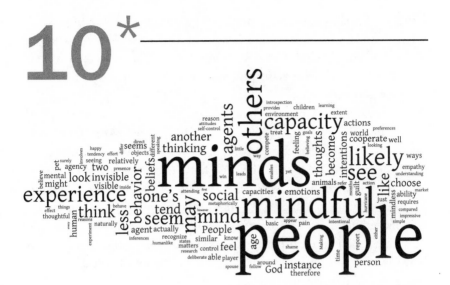

Seeing invisible minds

Shortly after taking off from LaGuardia Airport in the dead of winter, the engines of US Airways Flight 1549 failed after inhaling several large geese. The pilots glided their plane onto the Hudson River, where all the passengers were rescued—cold, wet, and almost completely unharmed. Explained one passenger, "God was certainly looking out for us." New

* The lead author is Nicholas Epley, Ph.D., a Professor of Behavioral Science at the University of Chicago Booth School of Business. His research investigates people's ability to reason about others' minds—from knowing how one is being judged by others to predicting others' attitudes, beliefs, and underlying motivations—and the implications of systematic mistakes in mind reading for everyday social interactions. His research has appeared in more than two dozen journals; has been featured by *The Wall Street Journal*, CNN, *Wired*, and National Public Radio, among many others; and has been funded by the National Science Foundation and the Templeton Foundation. Epley has written for the *New York Times*, produced lectures for the *Financial Times*, and been elected as a Fellow

Orleans Mayor Ray Nagin offered a very different assessment of God's mind following the devastating impact of Hurricane Katrina when he explained, "Surely God is mad at America. Surely he's not approving of us being in Iraq under false pretense. But surely he's upset at Black America, too."

Depending on your own beliefs, such statements will seem somewhere between insane and insightful. To psychologists, they seem impressive. They seem impressive not because they reveal a keen sense of causal inference, but rather because they reveal what may be the most impressive capacities of the social brain—the ability to reason about, or "to see," what other minds see. Introspection enables you to know your own intentions, report on your own thoughts, feel your own pain, and recognize when you are feeling shame rather than guilt. Other minds, however, are inherently invisible. You cannot know what it is like to be another person on the inside because your skull gets in the way.

The inherent invisibility of other minds poses a major problem for hard-nosed philosophers, who skeptically note that people cannot infer that other minds exist. Although it is surprisingly difficult for philosophers to reject the skeptical conclusion from the "other minds' problem," almost everyone else casts it aside sometime around the age of five. At this point, people have developed such a strong capacity to think about other minds that they not only see minds in other people, but they seem to see other minds almost everywhere.[1] Gods can be caring or callous. Pets can be thoughtful or devious. And every now and then, computers can have a mind of their own.

Having the capacity to reason about other minds enables people to form deep social connections with others, to empathize with others' pain and share in their joy, and to anticipate others' actions. But having a capacity and actually using it are two different things. People are not naturally inclined to see invisible things in the environment. People do not

of the Association for Psychological Science. He was the winner of the 2008 Theoretical Innovation Prize from the Society for Personality and Social Psychology.

Other minds are inherently invisible. You cannot see an attitude, smell a belief, or touch an intention, yet you can nevertheless "see" these mental states in other people with great ease. You can even see them in agents ranging from pets to gadgets to gods. How you are able to see other minds, and how they become visible, matters because it marks the difference between treating others as human beings worthy of moral care and concern versus treating others as objects or animals.

automatically see into other minds, either, but instead do so only under certain circumstances and with some psychological effort. This chapter describes how people come to see other minds, how other minds can become more or less visible, and why the visibility of other minds matters for everyday life.

Kinds of minds

Studying how people understand other minds first requires understanding how people intuitively define another mind. Research suggests that people think of minds as having two distinct dimensions: the ability to act (agency) and the ability to feel (experience).[2] Mindful agency involves the cognitive activities that enable action, such as the capacity to plan, to have intentions, to engage in deliberate self-control, and to pursue one's own goals. Conscious preferences, attitudes, and beliefs follow from these capacities. Mindful experience, in contrast, includes the cognitive activities involved in reacting to the external world, such as the capacity for self-awareness, and the experience of basic psychological states (such as hunger, thirst, or pain) and other-oriented emotions (such as empathy or sympathy). Mindful experience also involves the capacity for metacognition—the capacity to think about one's own thoughts or emotions—exemplified in experiences such as confidence and doubt, or in secondary emotions such as shame, guilt, joy, or hope. People seem to represent these two capacities in others quite independently. A sociopath, for instance, can appear to act with a high degree of mindful agency but no mindful experience, whereas a baby might appear full of mindful experience with relatively little mindful agency. Seeing other agents as mindful essentially means seeing them as able to consciously think or to feel. Because you are aware of both your own thoughts *and* feelings, you likely consider yourself—as most people do—to be very mindful.

Making other minds visible

Introspection provides a kind of flashlight that seems to offer direct access to one's own agency as well as experience. Although research demonstrates that introspection is actually a process of indirect inference, it *appears* to us that introspection provides direct access to the workings of our own brain. It seems that we can look on the inside and *feel* when we are suffering, *know* when we are experiencing regret, or be

aware of our intentions to lose weight. We can use our introspective abilities to make inferences about what others are likely thinking or feeling
(and we very often do), but such inferences are likely to seem inherently
less direct, less immediate, and less illuminating. We cannot see into
other minds as clearly as we seem to be able to see our own.

When something is difficult to see, people may doubt whether it
actually exists, or they may not see it. This is true of the relative difficulty
that people have seeing other minds compared to one's own, in ways that
are sometimes very subtle and surprising. For instance, we tend to evaluate ourselves by consulting our mindful intentions, but we evaluate
others (and *their* intentions) by observing their actions. We may consider
ourselves to be conscientious if we planned to buy our spouse a birthday
gift, but we need to see an *actual* gift to infer that our spouse is equally
conscientious.[3] We tend to believe that we are more likely than others to
experience complicated mental emotions such as shame, guilt, or embarrassment.[4, 5] We tend to believe that our own behavior will therefore be
guided by moral sentiments such as empathy, guilt, or compassion
whereas others' behavior is more likely driven by the relatively mindless
motives of self-interest.[6] Other minds are more opaque than our own,
and some learning, attending, seeking, and projecting is required for our
brains to become fully social and see into them. Here's how.[7]

Learning

Children do not enter the world able to think about other minds, but
they learn to do so fairly quickly. At around three months of age, children
start preferentially attending to animate objects compared to inanimate
objects; at around six months of age, they start to distinguish between
intentional (goal-directed) and unintentional (accidental) action. At this
age, for instance, children will look reliably longer at a person who is
reaching for a cup than to a person who is making the same reaching
motion in the absence of a cup. During the next 18 months, children
become more likely to mimic intentional than unintentional actions,[8] to
follow the gaze of another person and, therefore, share his or her attentional focus, and to recognize that other people may have preferences
that differ from one's own. By age two, children's social ability to read
other minds seems to have already surpassed that of our nearest primate
relatives. During the next two years, they pick up what appear to be

uniquely human mind-reading capacities. By age five, children demonstrate the most sophisticated of mind-reading abilities—the capacity to recognize that others' beliefs may differ from one's own and to use those differing (sometimes *mistaken*) beliefs to predict the other agent's behavior. Variability from age five onward comes from learning more specific details about how other minds actually work, largely gathered from personal experience, religious practice, or broader cultural norms.

Many psychologists and neuroscientists speculate that learning to read other minds comes from the deeply social tendency, present at birth, to mimic others' actions. For example, Steve Small examines a neurological underpinning for mimicry, and Jean Decety argues that an inborn capacity for mimicry underlies the human capacity for empathy. Looking where others look and copying their actions is a reasonable way to copy their likely mental states as well. This egocentric method of using one's own mental experience as a guide to other minds continues to be employed throughout adulthood. This method can give insight into others' mental states but can also lead people to overestimate the extent to which others' minds are similar to one's own. These biases tend to be called egocentric when reasoning about other people, and they tend to be called anthropomorphic when reasoning about nonpeople, such as a god, a gadget, or a pet.

Attending

People rarely notice things in their environment unless they are specifically attending to them. Other minds likewise tend to be relatively invisible unless attention is specifically drawn to them. For instance, take a moment to think about how happy you are compared to the average American. No, really, please take a moment. If you just spent some time thinking about how happy you are and no time thinking about how happy the average American is, then you are no different from the majority of people in psychology experiments who do likewise.[8] People can consider others' thoughts, feelings, beliefs, or emotions, but doing so requires mental effort that is in short and limited supply. Consider, for instance, a simple experiment in which you are playing a game with another person.[9] Both you and the other player in this game must first choose privately to cooperate or compete with each other. If you both choose to cooperate, you both earn $5. But if the other player chooses to cooperate and you choose to compete, then you win $10 and your partner gets nothing. However, if you choose to cooperate and the other player chooses to compete, you win nothing and your partner wins $10. If you both choose to compete, you both win a measly $2. It seems obvious in this situation that you should consider both what you would like to do, but also what the other player is likely to be thinking. Experimental evidence suggests that people do indeed think about the former but spend little time thinking of the latter. In a basic version of this experiment, 60% of people chose to cooperate. However, when people were simply asked to think about their partner's thoughts before making their choice, only 27% chose to cooperate. If people had already been thinking about others' thoughts, as it seems like they would naturally be doing in this situation, then a simple instruction to consider others' thoughts would likely have no effect on people's own behavior. That is not the case. This simple experiment shows that people may not naturally consider other minds even when it appears that they should. Thinking about other minds requires attentional effort. It does not necessarily come automatically. Indeed, as Tanya Luhrmann describes in Chapter 12, "How does God become real?" people may have to work very hard to discern other minds, such as the mind of God, even when they are actively looking for them.

Seeking

If seeing other minds requires attention, then other minds are especially likely to become visible when people are motivated to think about them. There are many reasons why people might try to get into the mind of another agent, from a spouse, to a pet to a god, but two are reasonably well supported by existing research. First, people tend to think about other minds when they are trying to form a social connection with others.[10] Making people feel lonely or isolated, for instance, increases the tendency to describe one's pet as thoughtful, considerate, or sympathetic (all mindful traits). And those who are made to feel lonely are also likely to report believing in mindful supernatural agents, such as God. Second, people tend to think about other minds when they are trying to achieve some understanding and control over their environment or over another agent's behavior.[1] Concepts of mind, including attitudes, beliefs, goals, or desires, can provide compelling cause-and-effect explanations of behavior that give a sense of predictability and control. A meteorologist may know that a hurricane could strike *here* or *there,* depending on environmental conditions. Lacking such knowledge, a hurricane that strikes here rather than there may lead people to invoke a mindful agent—such as God—to explain that action.

Projecting

If introspection makes your own mind visible, then you might assume that those who look similar to you on the outside might look similar to you on the inside as well. Indeed, animals that move at a humanlike speed (such as a horse) seem more mindful to people than do animals that move either much slower (such as a sloth) or much faster (such as a hummingbird).[11] And an agent that looks humanlike, such as a computerized avatar with a human face, seems more mindful than avatars that do not look humanlike. Any parent knows how well toy makers love to capitalize on this tendency. But the converse of this effect is even more interesting—other minds become less visible as the agent becomes less similar to the self. Others that differ from you in their interests, nationality, or social status are likely to be seen as less mindful—less thoughtful, less likely to experience complicated emotions, less able to experience pain or suffering—than those who are similar to you. It is therefore understandable that the history of human conflict is filled with instances of people dehumanizing radically different others, treating them as mindless animals or objects.[12]

Why minds matter

It may not be obvious to you why it matters that your neighbor occasionally thinks that her computer has a mind of its own, that a mayor believes that God punished his city by sending a hurricane, or that you truly believe that your pet poodle is thoughtful and considerate. Some of these attributions of mind seem purely metaphorical and therefore unimportant, as ways of speaking rather than ways of believing; others seem to represent genuine beliefs about the real presence or absence of another social mind. But metaphors can influence behavior in ways that are consistent with believing the metaphor is literally true. People metaphorically refer to feeling dirty after behaving unethically, yet washing one's hands actually reduces the guilt that people report feeling from engaging in unethical actions.[13] People are surely just speaking metaphorically when they refer to rejection as being given the cold shoulder, yet research demonstrates that people do indeed report feeling that a room is colder after someone has just rejected them than after someone has just accepted them.[14] And surely people are speaking only metaphorically when they refer to the stock market as anxious or jittery, yet describing the market as mindful leads stock traders to believe that trends are likely to continue, whereas describing the market as mindless leads traders to predict more random variability.[15] Whether metaphorical or literal, seeing other agents as mindful matters because people tend to treat the agent as if it has a mind. That matters for at least three major reasons.

First, mindful agents have both intentions and the capacity for self-control. Mindful agents can therefore be held responsible for their actions. Possessing a guilty mind *(mens rea)* is necessary for being held criminally responsible for a crime in the United States, a precedent found in courts around the world dating back to the Middle Ages. In distant times and places where people did not so naturally restrict intentional capacities to other humans, animals (such as rats) and objects (such as "possessed" statues) were common targets of criminal prosecution.[16, 17] Increasing the extent to which other agents seem mindful also increases the praise or blame that they receive for their actions. And even diminishing the extent to which people feel in mindful control of their own behavior (such as by undermining people's belief in free will) leads people to behave in ways that are consistent with diminished self-control (such as cheating on an exam when tempted to do so).[18]

Second, other minds are capable of thinking and may, therefore, be thinking about *you*.[19] Being under scrutiny by mindful agents has two basic effects on human behavior. One is that mindful agents become sources of social influence, increasing the extent to which people behave in socially desirable ways.[20] Imagine, for instance, how you might behave if you found a magic ring that made you invisible, and you'll get the point. This ability for mindful surveillance to control behavior has been proposed as one of the reasons, if not the primary reason, religious systems that posit an omnipresent deity are able to maintain such large-scale cooperative societies. The other effect of mindful surveillance is that it is mentally taxing to monitor others' thoughts. This effortful monitoring can diminish a person's performance on other cognitively demanding tasks.[21] And while waiting for a stressful event, such as giving a speech, people show fewer stress-related responses when in the presence of their relatively mindless pet than when in the presence of their relatively more mindful spouse.[22]

Finally, other minds matter because mindful agents become moral agents worthy of care and compassion.[7] The principle of autonomy captures this most basic of human rights: that because all people have the same minimal capacity to suffer, deliberate, and choose, no person can compromise the body, life, or freedom of another person. Agents with mindful experience—the capacity to suffer, deliberate, and choose—become those that evoke empathy and concern for well-being, whereas agents without mindful experience can be treated simply as mindless objects. From debates about abortion to animal rights to euthanasia, the mindful experience of the agents in question is often either the explicit or implicit focus of debate. Making invisible minds visible, and hence more like one's own, enables people to more readily follow the most famous of all ethical dictates: to treat others as you would have others treat you.

Conclusion

It is impossible for scientists to examine whether God was looking out for the passengers of Flight 1549 or punishing the residents of New Orleans with Hurricane Katrina, but it is very possible to examine why people might make such inferences. These examinations have revealed a remarkable capacity to look beyond the visible behavior that the environment provides to reason about a completely invisible world of intentions and goals, of motives and beliefs, of attitudes and preferences—an invisible world of other minds. Understanding when people are likely to recognize other minds, and when they are not, is the key to understanding when people may be likely to invoke natural versus supernatural explanations, when gadgets can seem to have minds of their own, and when people are likely to treat their pets as people and their enemies as animals. A mind like our own, with the capacity to see into other minds, is essential for an agent to be, as we are, fundamentally social.

Endnotes

1. N. Epley, A. Waytz, and J. T. Cacioppo, "On Seeing Human: A Three-Factor Theory of Anthropomorphism," *Psychological Review* 114 (2007): 864–886.

2. H. M. Gray, K. Gray, and D. M. Wegner, "Dimensions of Mind Perception," *Science* 315 (2007): 619.

3. J. Kruger, and T. Gilovich, "Actions, Intentions, and Trait Assessment: The Road to Self-Enhancement Is Paved with Good Intentions," *Personality and Social Psychology Bulletin* 30 (2004): 328–339.

4. J. P. Leyens, P. M. Paladino, R. T. Rodriguez, J. Vaes, S. Demoulin, A. P. Rodriguez, and R. Gaunt, "The Emotional Side of Prejudice: The Role of Secondary Emotions," *Personality and Social Psychology Review* 4 (2000): 186–197.

5. L. Van Boven, G. Loewenstein, and D. Dunning, "The Illusion of Courage in Social Predictions: Underestimating the Impact of Fear of Embarrassment on Other People," *Organizational Behavior and Human Decision Processes* 96 (2005): 130–141.

6. N. Epley and D. Dunning, "Feeling 'Holier Than Thou': Are Self-Serving Assessments Produced By Errors in Self or Social Prediction?" *Journal of Personality and Social Psychology* 79 (2000): 861–875.

7. N. Epley, and A. Waytz, "Mind Perception," in S. T. Fiske, D. T. Gilbert, and G. Lindsay (eds.), *The Handbook of Social Psychology* (New York: Wiley, in press).

8. J. Kruger, "Lake Wobegon Be Gone! The 'Below-Average Effect' and the Egocentric Nature of Comparative Ability Judgments," *Journal of Personality and Social Psychology* 77 (1999): 221–232.

9. N. Epley, E. M. Caruso, and M. H. Bazerman, "When Perspective Taking Increases Taking: Reactive Egoism in Social Interaction," *Journal of Personality and Social Psychology* 91 (2006): 872–889.

10. N. Epley, S. Akalis, A. Waytz, and J. T. Cacioppo, "Creating Social Connection through Inferential Reproduction: Loneliness and Perceived Agency in Gadgets, Gods, and Greyhounds," *Psychological Science* 19 (2008): 114–120.

11. C. K. Morewedge, J. Preston, and D. M. Wegner, "Timescale Bias in the Attribution of Mind," *Journal of Personality and Social Psychology* 93 (2007): 1–11.

12. N. Haslam and P. Bain, "Humanizing the Self: Moderators of the Attribution of Lesser Humanness to Others," *Personality and Social Psychology Bulletin* 33 (2007): 57–68.

13. C. B. Zhong and K. Liljenquist, "Washing Away Your Sins: Threatened Morality and Physical Cleansing," *Science* 313 (2006): 1451–1452.

14. C. B. Zhong and G. J. Leonardelli, "Cold and Lonely: Does Social Exclusion Literally Feel Cold?" *Psychological Science* 19 (2008): 838–842.

15. M. W. Morris, O. J. Sheldon, D. R. Ames, and M. J. Young, "Metaphors and the Market: Consequences and Preconditions of Agent and Object Metaphors in Stock Market Commentary," *Organizational Behavior and Human Decision Processes* 102 (2007): 174–192.

16. P. S. Berman, "Rats, Pigs, and Statues on Trial: The Creation of Cultural Narratives in the Prosecution of Animals and Inanimate Objects," *NYU Law Review* 69 (1994): 288–326.

17. C. R. Sunstein and M. C. Nussbaum, *Animal Rights: Current Debates and New Directions* (New York: Oxford University Press. 2004).

18. K. D. Vohs and J. W. Schooler, "The Value of Believing in Free Will: Encouraging a Belief in Determinism Increases Cheating," *Psychological Science* 19 (2008): 49–54.

19. A. Norenzayan and A. F. Shariff, "The Origin and Evolution of Religious Prosociality," *Science* 322 (2008): 58–62.

20. A. H. Buss, *Self-Consciousness and Social Anxiety* (San Francisco, Calif.: W. H. Freeman, 1980).

21. S. L. Beilock and T. H. Carr, "When High-Powered People Fail: Working Memory and 'Choking under Pressure' in Math," *Psychological Science* 16 (2005): 101–105.

22. K. M. Allen, J. Blascovich, and W. B. Mendes, "Cardiovascular Reactivity and the Presence of Pets, Friends, and Spouses: The Truth about Cats and Dogs." *Psychosomatic Medicine* 64 (2002): 727–739.

Inferring minds where
none can be seen

The social brain seeks connections with others. But what foundation do we use to build such connections? We experience empathy as a form of emotional resonance and understanding of other people. This connection allows us to comfort, support, and celebrate with others. Being in tune with emotional states of others allows us to respond in ways that strengthen a group. But how do we understand the thoughts and goals of others? How do we predict choices and decisions to facilitate cooperation in groups? Anthropomorphism is the basis for predicting behavior, thoughts, and goals. Nick Epley discusses how anthropomorphism is rooted in an egocentric view of others. Moreover, our view of others is not confined to the others that are people. It is, perhaps, reflective of the deep and fundamental nature of anthropomorphism in the social brain that its anthropomorphic inferences about agents can be derived from observed behavior, allowing us to understand "minds" where none may exist, as in mechanical toys or alarm clocks. Of course, we tend to understand those minds by thinking that they are like us.

Even when there is no agent to be seen, events in the world may be understood by attributing them to unseen agents. During World War II, the bombing of London was demonstrably random, but citizens of London could not help but discern intentional patterns in the attacks. As Epley points out, hurricanes and floods are even today attributed to the hand of God, perhaps an angry God. In Chapter 11, "Anthropomorphism: human connection to a universal society," Clark Gilpin discusses how religions may use this aspect of the social brain to achieve an understanding of God and what God wants. This kind of anthropomorphism can be taken to different metaphoric extremes in personifying God as father or friend. But an overly concrete personification may have costs, perhaps diminishing the universality and pervasiveness of God in other religions. Thus, religions may differ in theological perspective on the value of the anthropomorphic impulse inherent in the social brain.

Anthropomorphism: human connection to a universal society

When Jonathan Edwards, an angular New England minister in his late thirties, mounted the narrow steps into the pulpit on July 8, 1741, the sermon he was about to preach would become one of the most famously electrifying orations in American history, *Sinners in the Hands of an Angry God.* Edwards preached this sermon during the massive transatlantic religious revival that gave rise to Methodism in England and that

---

* The lead author, Clark Gilpin, Ph.D., is the Margaret E. Burton Professor of the History of Christianity at the University of Chicago Divinity School. Gilpin studies the cultural history of theology in England and America from the seventeenth century to the present. From 1990 to 2000, he served as Dean of the Divinity School, and from 2000 to 2004, he directed the Martin Marty Center, the Divinity School's institute for advanced research in all fields of the academic study of religion. His current research projects include a book with the working title *Alone with the Alone: Solitude in American Religious and Literary History*, which

came to be known in the American colonies as "the Great Awakening." This was not the familiar pulpit of his congregation in Northampton, Massachusetts, but rather the church at Enfield, a town that had gained notoriety for stubbornly resisting the exhortations of previous preachers of spiritual awakening. From his scriptural text—"their foot shall slide in due time" (Deuteronomy 32:35)—Edwards drew the doctrine that "There is nothing that keeps wicked men, at any one moment, out of hell, but the mere pleasure of God." Sinners living here and now, Edwards declared, were "the objects of that very same anger and wrath of God, that is expressed in the torments of hell," and that wrath was an annihilating fire that already "burns against them; their damnation does not slumber; the pit is prepared; the fire is made ready...to receive them." In a notorious image, Edwards portrayed God dangling the sinner's soul over the fires of hell like a spider on a single, slender filament of its web. The sermon achieved stunning results, as recorded in the diary of one of those present, Stephen Williams: "Before the sermon was done, there was a great moaning and crying out throughout the whole house—what shall I do to be saved; oh, I am going to hell; oh, what shall I do for a Christ." The "shrieks and cries were piercing and amazing," Williams reported, and the scene was so tumultuous that Edwards had to stop before finishing his sermon.[1]

As the classic American example of fire-and-brimstone Protestant preaching, *Sinners in the Hands of an Angry God* depends, for its effect, on *anthropomorphism*: ascribing human form and attributes—hands, emotions, and purposive agency—to nonhuman phenomena. Anthropomorphism has been a hotly debated feature of religion since classical

explores ways in which the spiritual discipline of solitary writing—autobiographic narratives, journals, and letters—shaped the careers of major New England intellectuals of the eighteenth and nineteenth centuries.

Anthropomorphic representations of God make many modern people very nervous, including many religious people. Attributing humanlike ideas and emotions to the comprehending powers of the universe not only seems out of step with modern science, but also a presumptuous confinement of the world within merely human needs and capacities. Yet the impulse to speak anthropomorphically about our "ultimate environment" has vigorously persisted into the modern age. Rather than dismissing anthropomorphism as an outmoded way of thinking, this chapter adopts a historical approach to rethink why anthropomorphism exhibits this perennial capacity to focus the human ethical imagination on our relations with, and obligations to, the universe within which we live.

antiquity. In the modern world, religious anthropomorphism has become especially controversial, but it is also becoming a crucial concept in modern theories about the very nature of religion.[2] Modern objections to anthropomorphism have taken two major forms. First, traditions within Judaism, Christianity, and Islam have long opposed the worship of "idols" and regarded anthropomorphism as a dangerous assault against genuine piety and properly theological understanding of existence. In the modern period, these religious objections against anthropomorphism have eventuated not only in sophisticated intellectual polemics against anthropomorphic concepts of God, but also in popular movements of iconoclasm, which protested against anthropomorphic representations of the divine and sometimes physically destroyed anthropomorphic images. Second, the rise and development of modern science has emphasized the regularity of the processes that structure the natural world. And even when these orderly processes were described as natural "laws"—a term with obvious anthropomorphic connotations—they have generally been understood in ways that are thoroughly impersonal and lacking any intrinsic purpose or design. Hence, modern science has generated numerous questions about religious interpretations of influence on the course of nature by divine ideas and purposes. Since the late nineteenth century, many instances of the so-called "warfare" between science and theology have turned on the issue of whether any scientific plausibility could be attached to concepts of a divine mind, purpose, or intention that guided or ordered the structure of the universe.[3] According to the philosopher Charles Taylor, the transition from perceptions of a divinely ordered, purposive universe to an "impersonal order" of nature marked a pivotal change that, especially since the eighteenth century, has shaped "a secular age" among the societies of the modern West.[4] In short, anthropomorphism has become not only a source of tension within religions, but also something of an impasse between religious and scientific interpretations of the universe. Nonetheless, anthropomorphic assumptions remain vigorously present in many of the modern forms of theology, spiritual practice, and religious art—a persistence that suggests the strength of the psychological and social functions performed by anthropomorphic representation.

In light of these longstanding controversies about religious anthropomorphism, the graphically anthropomorphic, spider-dangling deity of Edwards's sermon would seem to offer a good test case for understanding how anthropomorphic religious language works in the modern era. Such an understanding begins with one of the central themes of this book, namely the powerful human motivation to establish and maintain social connections. Anthropomorphism extends this drive for social connection beyond the boundaries of human societies by attributing human characteristics to nonhuman phenomena. In this way, anthropomorphic language incorporates human society in a web of ethical obligations that connect to the natural environment and, by imaginative extension, to the universe as a whole. Although the drive toward social connection is a general human trait, persons neither seek nor find satisfaction in a generalized sense of connection. Instead, satisfying social connections are sought and experienced in terms of the social norms and values of particular historical and cultural settings. Likewise, anthropomorphism, as an inferred social connection to the nonhuman, takes shape and becomes persuasive in terms of historically and culturally specific assumptions about society and social relations. This chapter steps back from *Sinners in the Hands of an Angry God* to describe how contemporary scholarship in social neuroscience and in the history of religions provides a fresh point of view on the workings of anthropomorphic perception. Then we test that interpretive model by reappraising Edwards's famous sermon in its historical context.

The boundary of the human

The line between the human and the nonhuman is perhaps the most consequential presupposition that any society, group, or individual adopts about life in the world. The way various cultures draw this line, between "us" and "the other," has shaped civilizations and their goals, as well as the norms of personal conduct and identity. Although concepts of the human have a long and contentious philosophical history, people in their everyday lives show remarkable consensus in the features they use to define "human." Central to this process of perceiving the human is a perception of mind in other agents, including the presence of goal-directed agency; emotions such as anger, guilt, or pride; a capacity for self-awareness; and free will. As Nick Epley demonstrated in the preceding chapter, the perception of these distinctively human traits—"seeing invisible minds"—is a psychological mechanism with tremendous influence on the way humans order and understand their social environment.

Mind perception is such a powerful tool of inductive inference, however, that it regularly crosses the line it has drawn between the human and the nonhuman. Scholars from a wide array of disciplines have long observed humans' anthropomorphic tendency to see nonhuman things or events as humanlike, imbuing the real or imagined behavior of nonhuman phenomena with human motivations, agency, and emotions. By perceiving the world in terms of human capacities and social relationships, anthropomorphism builds a complex system of analogies that uses knowledge of what it is like to be a person to interpret the behavior of animals, the function of technological devices, the operation of complex social systems such as "the market," or natural occurrences such as violent weather patterns or catastrophic events.[5] Hence, anthropomorphism, as a process of inference that not only draws but also crosses the line between the human and the nonhuman, has very substantial consequences for the human sense of connection to nonhuman animals, to larger ecological systems, and to the structure of the universe taken as a whole.

Contemporary psychological research has created an intellectual space that opens the phenomenon of anthropomorphism to fresh possibilities for interpretation that are especially important for understanding its role in the human spiritual traditions in modernity, as well as the controversies surrounding that role. This research proposes that a single set of psychological mechanisms is likely to explain when people perceive a mind at work in an encountered phenomenon, regardless of whether the thing in question is a god, a machine, an animal, another person, or an uncanny sequence of events. From this perspective, the psychological process of anthropomorphic inference works in concert with two other motivational mechanisms: the need to interact effectively with nonhuman phenomena in our environment, and the desire to establish social connections with other humans.[6] However, the phrase "anthropomorphic inference" fails to capture the interactive dynamism that infuses a person's perception that a mind is at work in another agent. Put more strongly, our sense that mind is present in the other is, in no small measure, the sense that we are *communicating*. The absence of this communicative dimension of mind perception is precisely the tragedy in the family of an Alzheimer's patient—the loss of reciprocal recognition. As Tanya Luhrmann vividly illustrates in Chapter 12, "How does God become real?" religious anthropomorphism builds on the notion that this communicative reciprocity extends beyond the boundary of human society into the wider environment, and includes social connection and communication with the divine.

Anthropomorphism adds an obvious but important twist to these psychological mechanisms: Whenever a person ascribes human attributes to a nonhuman phenomenon, the person nonetheless continues to perceive it as nonhuman. For example, when a pet owner observes a dog's reliability and infers that this behavior arises from the dog's faithfulness, the owner does not go on to say that the dog *is* human. Indeed, an indispensable aspect of an anthropomorphic way of seeing Fido's faithfulness is that the person also continues to see Fido as a dog. This dual perception is even more pronounced in a parallel illustration: A person observes the everyday reliability of gravity and infers that this arises from faithfulness at the heart of the natural order. Both illustrations indicate a close connection of anthropomorphism to metaphor, in which persons understand one kind of thing in terms of another by identifying a feature that bridges their difference without eliminating it. The specific feature—in this case, *faithfulness*—posits a point of comparison that enables a familiar human capacity for loyalty to enable interpretation of another, unfamiliar or alien phenomenon. The illustrations further suggest that Fido's faithfulness is what could be called *weak* anthropomorphism because one could plausibly argue that dogs and humans actually do share a capacity for faithfulness. We perceive that, as mammals, they exhibit many behavioral similarities. By contrast, the perception of faithfulness as an attribute of the natural order is *strong* anthropomorphism because it makes a far more daring inference in its effort to draw an analogy that produces insight or knowledge about the communicative reciprocity of the human and the nonhuman. As Edwards's sermon illustrates, religions build primarily on strong anthropomorphism to propound communicative social connection with the nonhuman world. The metaphorical and analogical reasoning characteristic of religious interpretations of the world is not merely a rhetorical flourish but is instead closely tied to general psychological processes in which self-knowledge and knowledge of other humans function as the most readily accessible starting points for inferences we make about human connections to the most comprehensive and consequential forces at work in the nonhuman world.[7]

Anthropomorphism's dual perception of nonhuman phenomena—as simultaneously both like and unlike human persons—has shaped modern religious and spiritual perceptions in two especially intriguing directions. In one case, it has fastened on the *difference* between the human and the divine and cultivated iconoclastic perceptions of the spiritual, in which anthropomorphic representation is regarded as a dangerously misleading, albeit necessary, accommodation to the limitations of human reason.[8] In the other case, it has emphasized the point of metaphorical *identity* between the human and the nonhuman. Thus, an ancient idea held that the physical universe was a macrocosm mirroring the human microcosm, and this included what literary critics have named "the pathetic fallacy"—a sympathetic response in nature to the affective states of humans. For example, the early modern political philosopher Hugo Grotius (1583–1645) described how human disobedience to divine law prompted an anthropomorphic emotional response from nature: "With sad motion wheeling, let the sky lament and mourn."[9] In the modern history of religions, these two modes of anthropomorphism—one accentuating difference and the other accentuating identity—have varied according to social circumstance, rhetorical purpose, and political ramifications. *Sinners in the Hands of an Angry God* illustrates, in one historical context, how anthropomorphic language crosses the boundary of the human to interpret human ethical responsibility to both the human and the nonhuman environment.

The rhetoric of divine wrath

Clearly, Edwards's anthropomorphic rhetoric destabilized in a terrifying way the Enfield congregation's complacent perceptions of the world. It did so by starting from an assumption that Edwards and the congregation shared: that humanity's ultimate environment should be construed, anthropomorphically, as a cosmic society held together by a covenant that God had made with the whole of creation. The sermon induced terror among the congregants by graphically portraying their own responsibility for disrupting the harmonious order of this all-encompassing society and provoking divine wrath for their rebellion against the covenant.

However, the primary evidence for this divine wrath did not come from the external orders of nature and divine providence, which Edwards imagined working together to maintain the harmonious order of the world. Instead, wrath had arisen from a clash between the benign will of the world and the rebellious human will: "There is laid in the very nature of carnal men a foundation for the torments of hell: There are those corrupt principles, in reigning power in them, and in full possession of them, that are seeds of hell fire." For the present, God restrained human wickedness "by his mighty power, as he does the raging waves of the troubled sea," but if God should withdraw that restraining power, humanity's willful self-regard would overturn nature. Therefore, the most dangerous fire in creation was not the fire of hell. Instead, it was the hellfire bursting forth from an unrestrained human will: "The corruption of the heart of man is a thing that is immoderate and boundless in its fury; and while wicked men live here, it is like fire pent up by God's restraints. But should God ever relax his governance, humanity's boundless fury "would set on fire the course of nature."

The turmoil stirred by human willfulness, like a violent storm at sea, threatened to capsize the ark of the universe. The Earth responded to this threat in a terrifying version of the pathetic fallacy, in which not empathy but enmity arose between humans and their natural environment. Consequently, except for "the sovereign pleasure of God, the earth would not bear you one moment," and Edwards admonished the Enfield congregation that "the creation groans with you" and resented its subservience to human usurpation:

> The sun don't willingly shine upon you to give you light to serve sin and Satan; the earth don't willingly yield her increase to satisfy your lusts; nor is it willingly a stage for your wickedness to be acted upon; the air don't willingly serve you for breath to maintain the flame of life in your vitals, while you spend your life in the service of God's enemies. God's creatures are good, and were made for men to serve God with, and do not willingly subserve to any other purpose, and groan when they are abused to purposes so directly contrary to their nature and end.

A rebellious humanity antagonized the rest of creation, and Edwards warned, "[T]he world would spew you out, were it not for the sovereign hand of him who hath subjected it in hope."

The just order of the cosmos would rightly destroy humanity for its willful rebellion against the order of the whole. The fact that this had not already happened was the expression of something like the self-restraining mercy of a monarch who does not order the execution of a traitor who has offended the royal honor: "The bow of God's wrath is bent, and the arrow made ready on the string, and Justice bends the arrow at your heart, and strains the bow, and it is nothing but the mere pleasure of God, and that of an angry God, without any promise or obligation at all, that keeps the arrow one moment from being made drunk with your blood."

The Great Awakening was coterminous and interactive with the eighteenth-century development of the modern physical sciences, especially building on the work of Isaac Newton (1642–1727). Edwards's assumptions about the harmonious order of creation combined the science of his day with the aristocratic social order of eighteenth-century society. In warning the town of Enfield that it had transgressed the cosmic order, Edwards was also asserting that it had violated the societal aspect of that order; as minister, he called the town to task for both violations.

Conclusion

Edwards imagined the Newtonian universe as an aristocratic social hierarchy held in harmony by sovereign law, both moral and natural. In this hierarchy, one member—the human—had stepped beyond its assigned place in the cosmic society and now lived in an unwitting complacency, ignoring the precarious finitude of a life being pursued by a radical judgment: "'Tis nothing but [God's] hand that holds you from falling into the fire every moment: 'Tis to be ascribed to nothing else, that you did not go to hell the last night; that you suffered to awake again in this world, after you closed your eyes to sleep. And there is no other reason to be given, why you have not dropped into hell since you arose in the morning, but that God's hand has held you up."

Like his contemporaries, the deists and religiously inclined scientists (such as Newton himself), Jonathan Edwards assumed the "Newtonian world machine," operating with the metronomic regularity of natural law. Presupposing both the science and the aristocratic social hierarchy of his day, Edwards introduced anthropomorphic language to create a clash between this harmonious order and the willful self-interest of humans who dared to ignore their proper rung on the ladder of existence. As a preacher of penitence, he carried his anthropomorphic imagery to extravagant heights to induce a reversal of behavior in a recalcitrant town. The sermon effectively threatened the people of Enfield with what amounted to "metaphysical ostracism," an expulsion no less thorough than the primordial ejection of Adam and Eve from the garden. The palpable effect of this imagery depended on the evocation of the natural and social orders rising up like, yet unlike, an angry monarch to crush rebels against the cosmic commonwealth.

Endnotes

1. In the interests of clarity, I have slightly rearranged and modernized this quotation. It and all quotations from Edwards's sermon are taken from Jonathan Edwards, *Sermons and Discourses, 1739–1742*, Harry S. Stout and Nathan O. Hatch, with Kyle P. Farley, ed., *The Works of Jonathan Edwards* (New Haven, Conn.: Yale University Press, 2003), 404-18.

2. Social scientific analysis of the relation between anthropomorphism and religion is summarized by Steward Elliot Guthrie, who argues "religion *is* anthropomorphism" in his book *Faces in the Clouds: A New Theory of Religion* (New York: Oxford University Press, 1993), 178. The most lucid and succinct historical treatment remains Frank E. Manuel, *The Eighteenth Century Confronts the Gods* (Cambridge, Mass.: Harvard University Press, 1959).

3. I have in mind such authors as John William Draper, *History of the Conflict between Religion and Science* (New York: D. Appleton and Company, 1874) and Andrew Dickson White, *A History of the Warfare of Science with Theology in Christendom* (New York: D. Appleton and Company, 1896). For a more recent example, see Richard Dawkins, *The God Delusion* (Boston: Houghton Mifflin, 2008).

4. Charles Taylor, *A Secular Age* (Cambridge, Mass.: Harvard University Press, 2007), 270-95.

5. For a sampling of relevant recent work, see Philip Husbands, Owen Holland, and Michael Wheeler, eds., *The Mechanical Mind in History* (Cambridge, Mass.: MIT Press, 2008); and Lorraine Daston and Gregg Mitman, eds., *Thinking with Animals: New Perspectives on Anthropomorphism* (New York: Columbia University Press, 2005). Ralph Waldo Emerson made the classic American argument for the positive reciprocity between the human and the natural. This idea of mutuality takes a different turn in our contemporary situation, in which industrial and

chapter 11 · anthropomorphism: human connection
to a universal society
143

technological advances have begun to alter the climate and thereby blurred the boundary between the human and the natural in another way. For examples of this phenomenon, see Bill McKibben, *The End of Nature* (New York: Random House, 1989).

6. Nicholas Epley, Adam Waytz, and John Cacioppo, "On Seeing Human: A Three-Factor Theory of Anthropomorphism," *Psychological Review* 114 (2007): 864–886.

7. Sheldon Sacks, ed., *On Metaphor* (Chicago: University of Chicago Press, 1978); David Tracy, *The Analogical Imagination: Christian Theology and the Culture of Pluralism* (New York: Crossroad, 1981).

8. Alain Besançon, *The Forbidden Image: An Intellectual History of Iconoclasm*, trans. Jane Marie Todd (Chicago: University of Chicago Press, 2000); Joseph Leo Korner, *The Reformation of the Image* (Chicago: University of Chicago Press, 2004).

9. Merritt Y. Hughes, "Earth Felt the Wound," *English Literary History* 36 (1969): 193–214.

Personifications of God

Jonathan Edwards wrote during a time in which monarchs reigned supreme. Writing from this perspective, he argued that the universe is a cosmic society organized under the leadership of a King of kings, a society against which humans have rebelled and, as a consequence, humans are at risk of annihilation except for the mercy of the King. Judgment day will come, according to Edwards, and those who have failed to meet their moral responsibility to the directive of the universe face eternal isolation. Clark Gilpin notes that by conjuring up a personified God—a God with emotions, intentions, and the capacity to act—Edwards instilled great fear and trembling in his listeners that presumably motivated them to change their behaviors in the desired direction. It is hard to imagine that Edwards would have had comparable success had he resorted to simple instructions or exhortations to engage in certain behaviors and avoid others. The innate tendency of people to understand divine entities in terms of what people do understand (namely their own thoughts, feelings, and beliefs) provided the leverage on which Edwards relied to drive his message home.

Tanya Luhrmann also discusses a personalized construction of God—a God with whom one can consult and who intervenes in one's daily life. People are intrinsically motivated to form social connections, and very little in life is more rewarding to people than their social relationships. Luhrmann, who adopts the perspective of a participant observer, finds that a new evangelical Protestant movement, the Vineyard Church, appeals to this motivation to depict God as one's personal guide and friend, well within one's sensory reach. By conjuring up an anthropomorphic God with loving emotions, intentions, and actions, the Vineyard Church creates a desire for a personal relationship with God. But developing a relationship with an invisible God defies rationality. People must learn how to transform an abstract concept of an invisible God into a concrete sensory presence in their lives. Just as the social brain can perceive nonhuman objects as human, the social brain is also capable of selectively attending to sensory experiences and interpreting these sensations as God's presence.

12 *

How does God become real?

How does God become real to people when God is understood to be invisible and immaterial, as God is within the Christian tradition? This is not the question of whether God is real, but rather how people learn to make the judgment that God is present. Such a God is not accessible to the senses. When you talk to that God, you can neither see his face nor

* The lead author is Tanya Luhrmann, Ph.D., a professor in the Stanford Anthropology Department. Her interests include the social shaping of psychological experience, and the way social practice may affect even the most concrete ways in which people experience their world, particularly in the domain of what some would call the "irrational." Her first project, *Persuasions of the Witch's Craft* (Harvard, 1989), was a detailed study of the way apparently reasonable people come to believe apparently unreasonable beliefs. Her second project, *The Good Parsi* (Harvard, 1996), explored the apparently irrational self-criticism of a postcolonial India elite, the result of colonial identification with the colonizers. Her third book, *Of Two Minds* (Knopf, 2000), identified two cultures within the

hear his voice. You cannot touch him. How can you be confident that he is there?

Anthropology cannot answer the question of whether God is real. But the traditional method of the discipline, participant observation, can use the slow, careful method of fieldwork to explore the way people learn to experience God as present in their lives. And what the method can teach us is that this often intensely private and personal relationship between a creature and its creator is built through a profoundly social process.

In fact, one of anthropology's most useful contributions to understanding the experience of God is to draw attention to just how much work faith takes, and to the fact that different kinds of faith—and different understandings of God—demand different kinds of work. Many who do not believe in God approach the question of religious belief as the problem of why people should believe in the existence of an invisible, intentional agent in their world. One of the more persuasive recent answers emerges from the observation that many of our cognitive traits evolved to help us survive. Evolutionary anthropologists and psychologists argue that belief in invisible beings is an accidental byproduct of the way our minds have evolved over the millennia. Our quick, effortless, automatic intuitions lead us to "anthropomorphize," or to see faces in the clouds, as one scholar puts it. From this perspective, people believe in God because it is so easy to believe in invisible supernatural presence, and the great religions are elaborations around this basic core.[1]

American profession of psychiatry and examined the way these different cultures encouraged two different forms of empathy and understandings of mental illness. She trained at the University of Cambridge (Ph.D. 1986), and taught for many years at the University of California San Diego. Prior to moving to Stanford, she was the Max Palevsky Professor and a Director of the Clinical Ethnography project in the Department of Comparative Human Development at the University of Chicago.

Religion is often understood as a matter of belief: a yes/no proposition. This chapter suggests that it may be more helpful, and more accurate, to understand religious commitment as a response to sensory experience that can be learned, and that the capacity to learn depends upon one's knowledge and belief, one's proclivity for experiencing the world in particular ways, and the impact of devotional practice.

Belief in the invisible

Yet it is also true that, in many ways, it is hard for people to believe in the invisible, intentional being of God, at least in some ways and at some times. It is one thing to believe in the abstract that there is a good and loving God; it is another thing to believe that this God loves you in particular this very afternoon when your car has broken down in the rain. Many Christians struggle at some point with whether God exists or whether they understand God's nature. A young man may come to university as a devout Christian, take a course on religion, and begin to wonder whether Christ and Krishna are cultural constructions. A depressed woman may understand herself as devout but find that, when she sits down to pray, she feels that no one is listening to her prayers. And always there are times when terrible things happen to good and faithful people who often continue to believe in God in the abstract, but who find that they can no longer pray at all. The struggle between espoused religion (the religion one asserts, such as through the Nicene creed) and lived religion (the way in which one experiences God from moment to moment) is central to the life of the Christian, and perhaps to the lives of most believers.

The problem for believers is that, to experience the Christian God as present, one must override three basic features of human psychology, features that are also part of our evolutionary inheritance. A person must override the expectation that our minds are private, an expectation so substantial that researchers have shown that it develops around the world at a more or less similar age and can be found even in nonhuman primates. A person must override the expectation that persons are visible. Finally, a person must override the expectation that love is conditional, as it is for all social beings beyond a certain age, when right behavior is expected as a condition of human interaction. At least some versions of Christianity expect unconditional love.

The deep puzzle of faith is not *why,* but *how.* How is it possible that people are able to violate such fundamental expectations of presence? The answer, in part, is that they do not. For most Christians, it will be a lifelong process to believe in all times and in all ways that their God is real for them in the way that their church tells them God is real. As the psalmist laments, "How long wilt thou forget me, O Lord? For ever? How long wilt thou hide thy face from me?" (Psalm 13:1). What they do to make God plausible for them depends upon their understanding of God and on what the social world of a faith teaches about how to experience their minds and bodies to find evidence of God's presence.

Learning to sense the presence of God

In 2004, I set out to study ethnographically the way God becomes real for people in a church that would exacerbate the cognitive burden of belief. I chose an example of the new Protestant church that grew up after the 1960s.[2] Those churches set out as an invitation to experience God as concretely and as vividly as God had been experienced by the earliest Christian disciples. This God is both intensely human and intensely supernatural. In these churches, God is understood as so personlike that he becomes someone to joke and argue with, or someone to chat with when walking down the street about trivial things that matter only to the congregant. Coming to know God in such a church is described as hearing God "speak." Dallas Willard, a beloved evangelical's intellectual, puts it baldly: that God's face-to-face conversations with Moses are the "normal human life God intended for us." I conducted ethnographic fieldwork at a church that exemplifies this approach to God, a Vineyard Christian Fellowship in Chicago and then on the San Francisco peninsula. (There were eight such churches in Chicago and four on the peninsula.) For three years I went to Sunday morning gatherings. I joined three small groups, or housegroups, each for a year; I went to conferences and retreats; and I interviewed many congregants casually and also more formally about the way they experienced God.

Overall, I observed that the process of coming to know God in such a church could best be described as a mapping process in which the congregants learn to use the familiar experience of their own minds and bodies to give content to the abstract experience of God. This is the way humans learn most commonplace abstract words, in effect cognitively mapping from what we know to what we can only imagine. God speaks, so congregants learn to infer from their own experience of inner speech the way in which God talks to them. God relates, so congregants learn to imagine a relationship with God based on their own experience of relationship. And God loves, so congregants use their own experience of being loved by a human as an example of the way they are loved by God. But unlike learning about time, congregants also map back. They build up a model of God by interpreting out of their own familiar experience into a representation shaped by the social world of the church and the narrative of the sacred text; then they seek to remap their own interior emotional experience by matching it to this representation. This demands constant effort, continual work on the way one pays attention and interprets one's experience. As an ethnographer, I could see three kinds of work.

First, God must be recognized as present. What congregants learn to do is to cherry-pick mental events out of the everyday flow of their awareness and to identify those moments as other than themselves, as being of God. God was said to speak in several different ways. He spoke through the Bible, so a verse "jumped out" at a congregant or in some way drew the attention. For example:

> I was reading in [some book] and I don't even know why I was reading it. There's a part where God talks about raising up elders in the church to pray for the church. And I remember, it just stuck in my head and I knew that the verse was really important and that it was applicable to me. I didn't know why. It was one of those, let me put it in my pocket and figure it out later." How did she know that it was important? "Because I just felt it. I just felt like it really spoke to me. I don't really know why. And a couple of days later a friend asked me to be on the prayer team and it was like, wow, that's what it was.

God spoke also through circumstances. What a skeptic might interpret as coincidence is understood as God's intentional decision to direct the congregant's attention. For example:

> Everything in my life right now is focused on trying to get to England, and I needed to get some ID pictures. So I was really anxious—the money hasn't really come together—and one afternoon I just felt like God said, you need to get up and go get those. Go get those ID pictures that you need. I was like, that's totally inefficient. I don't have a car, so it's like walking half an hour to Walgreens and another half an hour back. Like, I could do this later and combine it with several things I need to get done. But I felt it was a step of faith to do this thing. So I did it—grumbling. Then on the way there and back I ran into three people I knew, and I felt that there was a kind of pattern, and that I was in the right place at the right time.

These ways of recognizing God are widely shared in many forms of Christianity. More specific to experiential evangelical Christianity is the expectation that God will speak directly into the mind by placing a mental image or thought or sensation there. For example:

> I'm praying for someone and, you know, they say their situation, what they want me to pray for. I start praying and start trying to, you know, really experience God, and, you know, I see these vivid images, and I'm explaining these vivid images and what I think they mean and, you know, sort of checking in with the person, you know, does this resonate with you? They're like 'Oh, my gosh, yes! How did you know that?'

Most congregants find this process of pulling out specific thoughts and ascribing them to God baffling at first: Again, the process violates the basic human experience that the mind is private. A congregant commented, "Now I know that the 'something' is God, God's voice. But I didn't at all have words to describe it at that time. I didn't understand. It was very confusing."

The social world of the church taught specific ways to differentiate between mental events that are God and those that are not. This technique has been taught in the church since the earliest time as "discernment," although the content of the word and its rules has varied by era. In the modern experiential evangelical church, the rules of discernment are more often taught by example and gossip than explicitly. Nevertheless, there appeared to be four principles. A thought might be said to come from God if the thought was unexpected, the thought was consonant with God's nature, the congregant had additional confirmation (one "tested" the thought), and one felt peace during the experience. The process was understood to be ambiguous and left room not only for the congregant to be wrong, but for different congregants to disagree about whether God had, in fact, spoken in a particular manner. One afternoon, a woman spoke in front of the church explaining that God had spoken to her and told her that she should carry out some mission work in a lovely part of Mexico. The man sitting next to me said dryly, "God sure wants a lot of evangelizing in Puerto Vallarta."

Second, God must be experienced in relationship. Such churches invite congregants to experience God in their imaginations as a person. Again, this violates a basic psychological expectation: Persons have faces to observe and hands to shake. Human relational interactions are based on sensorial response. Churches like the Vineyard explicitly suggest that one should imagine a sensorial response from God and encourage congregants to participate in a kind of "let's pretend" play in which God was present. The pastor suggested one Sunday morning that congregants put out a second cup of coffee for God and sit down with him to chat. People went on a "date night" with God, to get a sandwich and sit down on park bench to talk with God as they imagined his arm around their shoulders. They would ask God truly trivial questions, like what shirt they should wear in the morning and what movie he thought they would like. These behaviors were clearly playlike. One congregant remarked, "I definitely do that. When I can't decide what to wear [I ask]… 'God, what should I wear?'" Then she laughed. "And you know, then I kind of forget about the fact that I asked God. I think God cares about really, really little things in my life. I mean, I know God cares, but I don't expect him to tell me what to wear. I'm like, Oh, I think I'll wear that and forget I even asked God!" This invitation to play was C. S. Lewis's explicit contribution to twentieth century Christianity: "Let us pretend to turn the pretense into a reality." In churches that encourage such play, heresy fades in importance. The pastors and the committed congregants worry about "deadness," not flawed imagining.

Third, congregants must learn to respond emotionally to God as if God is real. If God is real, a Christian (at least, the modern evangelical Christian) should experience the emotions that one would feel if one were loved unconditionally. Most do not. It is, in fact, difficult for humans to experience themselves as unconditionally loved because no matter how warm and loving a parent may be, at some point the child is expected to control his or her behavior and parental love becomes contingent. The task of feeling unconditionally loved imposes upon the congregant not only the burden of identifying and relating to an invisible being, but also the ability to experience emotions in response to that being's love that the congregant rarely, if ever, truly experiences. Congregants talk about the experience of unconditional love as rare: They speak of "those moments" when one really feels God's love.

> I was driving home from grocery shopping in the car and I stopped at a light, and suddenly for no reason that I could come up with, I was weeping and I felt a massive and awesome

sense of the presence of God in the car with me. It just came and I had absolutely no control over it. I pulled over to the side of the road—I remember thinking that I was *so in love* with Jesus at that moment that no one else on the planet could come close. After about twenty minutes of real intensity the feeling subsided somewhat, but the presence of Jesus stayed with me. I drove home not really ever able to fully express what happened without sounding like I'd taken something illegal.

The more immediate aim seemed to be to experience what Galatians 5:22–23 calls the "fruits of the spirit": love, joy, peace, patience, and so forth. The social life of the church was rich in emotional practices that sought to reshape the congregants' interior emotional world by modeling the self on God, or on the self as seen from God's loving perspective. One of the most important was prayer ministry, in which the person for whom the prayer is given is often crying and in visible pain. Those around the person offer prayers that describe the ways in which the sobbing person is loved by God. Another was to treat prayer like a psychotherapy session. One congregant explained, "It's just like talking to a therapist, especially in the beginning, when you're revealing things that are deep in your heart and deep in your soul, the things that have been pushed down and denied." In these churches that emphasize God's love and intimacy, hell and fear largely disappear.

The central demand of these learning practices is to use one's own mental experience as evidence and content for the responsive presence of this God, who is believed to be other than oneself, and to use pretend play to integrate those mental events into a representation that is persuasively external to the self. The emotional practices provide both direct evidence of God's love and, more generally, evidence that participation in church is satisfying and worthwhile. In effect, the process asks the congregants to carve God out of their own experience and to experience those phenomena as other, and it uses the emotional practices taught by the church and the social world of the congregation to help them hold that God separate and apart and lovingly responsive.

This is hard work to do, and not everyone was able to do it—or to do it easily. Consider two congregants describing their difficulty in experiencing God directly despite their efforts.

> Jake: I remember desperately wanting to draw closer to God, and [to have] one of these inspired Holy Spirit moments…. I wanted those [experiences] and I sought them out, but I never found myself encountering them.

> Irene: I don't understand the gift of prophecy completely. I probably never will, and I don't have it and I don't want it because it would scare me.

Another congregant describes her vivid experience of God:

> Nora: It was pretty early on in my relationship with him. I was just all full of myself one morning. I just had wonderful devotions and worships and just felt so close. I went out, and it was the most god-awful day. It was icy rain and gray and cold and it was sleeting. I'm just full of the joy of the Lord, and I say, 'God, I praise you that it isn't snowing, and that nothing's accumulating, and that the streets aren't icy'—and then I went around the corner, and I hit a patch of ice, and just about went down. It was so funny to me. I just burst out laughing out loud. It was just so funny that he would put me in my place in such a slapstick personal kind of way. But then he just graced me the rest of the morning. The bus showed up right away, which it never does. I was reading, and I missed my stop to get off, and I heard God say, 'Get off the bus.' I looked up and hollered, and the bus actually stopped, half a block on, to let me get off. I just felt that intimacy all morning. Like when you go from holding a new boyfriend's hand to kissing him goodnight….

Some people experience God speaking directly to them in an easy relationship. Others do not.

As a result of my involvement with the Templeton group, I decided to carry out some quantitative and experimental work to see whether we could figure out the differences between those who found it easy to do this work and those for whom it was difficult. That work suggests a psychological capacity that makes the process of knowing this kind of God easier, though its absence does not prevent religious experience and its presence does not predict it. It is the capacity for absorption, which is at the heart of imagination. Absorption is the capacity to focus one's attention on a non-instrumental (and often internal) object while disattending to everyday exterior surrounds. Absorption is related to hypnosis and dissociation but is not identical to either. All of us go into light absorption states when we settle into a book and let the story carry us away. There are no known physiological markers of an absorption state, but as the absorption grows deeper, the person becomes more difficult to distract, and his sense of time and agency begins to shift. He lives within his imagination more, whether that be simple mindfulness or elaborate fantasy, and he feels that the experience happens to him, that he is a bystander to his own awareness, more himself than ever before, or perhaps absent, but, in any case, different. And as the absorption grows deeper, people often experience more imagery and more sensory phenomena, sometimes with hallucinatory vividness. Scholars do not discuss training in absorption, although researchers of hypnosis and dissociation are clear that some kind of practice effects can be seen. [3]

Conclusion

Prayer is basically training in absorption, at least the kind of prayer in which the person praying focuses inwardly and disattends to the everyday world to engage with God. It would be hard to overestimate the importance placed on prayer and prayer experience in a church like this and, indeed, in Christian America today. Many of the best-selling Christian books are books on prayer technique, and they sell in the millions. Such books often begin by presenting the concrete sensory experience of God described in the Hebrew Bible as the everyday relationship for which the ordinary believer should strive. In these manuals, the act of praying is understood as a skill that has to be deliberately learned. I discovered that these evangelical congregants assumed that prayer was a skill that had to be taught, that it was hard, that not everyone was good at it, and that those who were naturally good and well trained would experience changes associated with a more richly developed inner world. Their mental images would seem sharper; they would be more likely to report unusual sensory experiences. In short, they would be more able to experience God. The more quantitative work—done in collaboration with Howard Nusbaum and Ron Thisted—suggests that those who have a proclivity for absorption and who trained that proclivity through prayer are indeed more able to accomplish the demanding learning that this concept of God sets out.[4] They are more able to identify God's presence in their mind. They are more likely to experience God as an invisible companion. They may be more capable of responding to God emotionally.

All theologies have trade-offs. This one offers an intensely personal and personlike God. He can comfort, like a friend, and respond directly, like a friend. He can be like a real social relationship for those who make the effort to experience him in this way. But because that social relationship lacks so many features of actual human sociality—no visible body, no responsive face, no spoken voice—such a theology demands a great deal of effort from those who follow it. They must constantly work with their attention, reinterpreting the ordinary and natural into the presence of the extraordinary and supernatural. Faiths that manage God differently— less personal, more present in the everyday natural world—make fewer demands on their followers' attentional habits. But perhaps such a God may be easier to take for granted. Paradoxically, this high-maintenance, effortful God may appeal to so many modern people (as many as a quarter of all Americans, according to a recent Pew study) precisely because the work demanded makes God feel more real in a world in which disbelief is such a real social option.

Endnotes

1. Scholars who contribute to this perspective include Scott Atran, Justin Barrett, Pascal Boyer, Stewart Guthrie, and Harvey Whitehouse.

2. These churches have been described by D. Miller, *Reinventing American Protestantism* (Berkeley, Calif.: University of California,1997); see also R. Wuthnow, *After Heaven: Spirituality in America Since the 1950s* (Berkeley, Calif.: University of California Press, 1998). A survey by the Pew Foundation (Pew, *Spirit and Power: Ten Nation Survey*, Pew Forum on Religion and Public Life, 2006) found that 23% of all Americans belong to a loosely similar style of "renewalist Christianity."

3. Good summaries of work on hypnosis and dissociation, with some reference to absorption, can be found in H. Spiegel and D. Spiegel, *Trance and Treatment* (New York: Basic Books, 2004); R. Seligman and L. Kirmayer, "Dissociative Experience and Cultural Neuroscience," *Culture, Medicine and Psychiatry* 32, no. 1 (2008): 31–64; and L. Butler, "Normative Dissociation," *Psychiatric Clinics of North America* 29 (2006): 45–62.

4. The empirical work is presented in T. Luhrmann, H. Nusbaum, and R. Thisted, "The Absorption Hypothesis," *American Anthropologist* (March 2010); and A. Tellegen and G. Atkinson, "Openness to Absorption and Self Altering Experiences ("Absorption"), a Trait Related to Hypnotic Susceptibility," *Journal of Abnormal Psychology* 83 (1974): 268–277.

Belief and connection

We tend to think of beliefs as wisps of the mind that have no power in the material world. However, as Gary Berntson and Louise Hawkley have discussed, beliefs can affect our health even to the extent of determining life and death. As Tanya Luhrmann discusses, in some forms of Christianity, there is a real belief in the presence of God. This is not simply a belief of God in the world, but a belief of a God who is by one's side. The idea of God as a friend and companion clearly motivates the desire to make such a presence manifest in tangible ways. For some, it is the sense of God with which they commune; for others, it is what they believe to be a sensory experience of God that they seek. Luhrmann outlines how this belief, coupled with a supportive social structure, can lead to powerful personal experiences, such as hearing the voice of God, reflecting the operation of our social brains.

Our sense of social connection is not dependent on a single set of religious beliefs, however. In human social connections, we can form individual relationships with a spouse or friends, but as John Cacioppo outlined, our social brain seeks other kinds of connections as well. We seek connections with emergent structures such as groups, clubs, teams, congregations, and beyond. Kathryn Tanner argues that the belief that God created the world and bears causal responsibility for it connects believers to life in a broader way than is provided through individual relationships. This broader connection to life does not depend on the manifestation of a presence to whom we can talk because the evidence of social connection is apparent in the very fabric of daily existence. Thus, whereas Luhrmann discusses God as a palpable friend that one can learn to attend to and experience as an active presence in one's life, Tanner discusses God as the initiator of life and the very fabric of existence, a presence so ubiquitous that there is no specific point on which one can focus to attend to or experience God.

13*

Theological perspectives on God
as an invisible force

An individual's beliefs about God are one factor to be included in a multidimensional investigation of the social consequences and possible health benefits of religion, an aid in particular to scientific hypothesis generation.[1] Scientists can better test for the social and health consequences of religious commitment when they know more about the character and range of beliefs about God that such commitment brings. This

---

* The lead author is Kathryn Tanner, Ph.D., the Dorothy Grant Maclear Professor of Theology at the University of Chicago Divinity School. Her research relates the history of Christian thought to areas of contemporary theological concern using critical, social, and feminist theory, with a special focus on the possible practical implications of Christian beliefs and symbols. She has lectured widely throughout the United States and Europe, and is the author of six books: *God and Creation in Christian Theology: Tyranny or Empowerment?* (Blackwell,

chapter hopes to show, in particular, that exactly what Christians believe about the nature of God's influence on their lives is likely to have an important bearing on one of the questions of this book: How can religion encourage a sense of connection to others, especially in situations of perceived social isolation, and thereby assuage the adverse health consequences of loneliness?

Depending on what they think God is like, Christians vary in the way they expect God to be a present influence on their daily lives. God's nature is supernatural or transcendent, which means God is not very much like any of the ordinary persons or things with which we come into regular contact. Christians use the same terms for God that they use for talking about ordinary persons and things, but they know that neither set of terms is really adequate to capture who or what God is. On one hand, God is something like a human being, in that God loves people and wants to do them good, and in that God is unhappy with their failings and trying, through the use of carrots and sticks, to get them to change. But on the other hand, God is really not much like an ordinary human being: God is present at all times and everywhere, working inexorably to bring about what He intends throughout the entirety of people's lives by way of both personal and impersonal influences—for example, through personal words of advice and warning found in the Bible, as well as apparent accidents of fortune such as car crashes and the weather. Though retaining personal characteristics such as love or anger, God operates less like an individual human person with limited reach and partial interests, and more like light, air, or gravity—quite pervasively and constantly.

1988), *The Politics of God* (Fortress, 1992), *Theories of Culture: A New Agenda for Theology* (Fortress, 1997), *Jesus, Humanity and the Trinity* (Continuum and Fortress Press, 2001), *Economy of Grace* (Fortress, 2005), and *Christ the Key* (Cambridge, 2010).

Christian beliefs are not just theoretical matters, involving putative truth claims about the nature of ultimate reality, but also practical ones: Christian beliefs are often promulgated with the hope of impacting the way human beings live, by establishing, for example, the meaningfulness of and motivations for certain forms of social behavior. Prior research has concerned the possible economic, social, and political consequences of Christian beliefs about God's relation to the world. This essay extends such questioning to the topic of perceived social isolation. How might belief in God as an invisible force in everyday life affect an individual's sense of social connection?

Usually one side comes to the fore in the way Christians feel connected to God. For some Christians, the personal side of God is central; for others, the more impersonal is. Thus, some Christians expect God to be much like a human friend, offering companionship and good advice.[2] As Tanya Luhrmann explores in Chapter 12, "How does God become real?" their religious lives often revolve around the internal sensory or imaginative experiences that make a God of that sort seem real to them—the sense that God is in the room with them, that God speaks to them, and so on. They work to cultivate a prayer life that heightens the vividness of those experiences and thereby allays doubts about the actual existence of this invisible, otherwise seemingly unreal, divine friend. In short, good practitioners of prayer gain a stronger and more reliable sense that God is present as a friend directing the course of their lives.

Other Christians have expectations of a more overarching and impersonal sort about the way God is a force in their lives. These expectations, I suggest shortly, are the consequence of their holding certain beliefs about God as the creator, sustainer, and savior of the world. In this case, a strong sense of God's presence in one's life does not depend on having experiences of a literally personal sort or on developing the spiritual practices that help cultivate such experiences. The sense that God is present as an influence on one's life is rather something one feels all the time simply by virtue of the beliefs one holds about God and the world. Given a particular construal of those beliefs—a relatively impersonal one, I argue—simply having those beliefs in mind, with some awareness of their quite obvious presuppositions and implications, makes clear one is never alone, never a self-sufficient operator. In contrast to the understanding of God as friend, here God's invisibility does not threaten to interrupt a sense of God's presence and influence. To the contrary, God's invisibility enables the sense of His presence and influence to be the routine backdrop of all one's experience, to constitute a general outlook on the world, no matter what the circumstances.

Belief in creation as a backdrop to the whole of life

For example, a common Christian construal of the belief that God is the creator of the world makes it possible for the sense that God is with one—as one's supporter and sustainer—to be a constant feature of one's life as a whole. Despite first impressions, God's creation of the world need not refer to the origin of the universe or to the beginning of its more specific features or components. Were either to be its meaning, belief in God as the creator of the world could not be a central component of a generally applicable Christian outlook or have much relevance for more than the occasional speculation about origins (for example, "Why am I here at all?"). The belief concerns instead a causal dependence upon God of a more continuous sort, spanning, in short, the whole time of the universe's existence and, therefore, the whole time of one's individual life. To be created by God is to exist in a relationship of dependence upon God for what one is, however, long one exists.[3] A human being therefore depends upon God for more than the fact of his or her birth: He or she remains dependent upon God in the same way ever after. God's creation of the world in general is simply not temporally indexed; it is no more closely associated with the beginning of things than with what comes later. A preoccupation with temporal origins therefore commonly drops out of Christian accounts of creation: The world is just as dependent upon God for its existence whether it has a beginning or always exists.[4] Belief in God's creation of the world for these reasons blurs into belief in God's supportive maintenance of it at every point in time.

Also enabling belief in God as creator to form a general backdrop to all one's experience—to be relevant on every occasion as a universally applicable worldview—is the fact that God is thought to be responsible as creator for the whole of what happens in the world at any one time. To believe that God is the creator of the world is, at the very least, to believe that God holds into existence the entirety of the world in any and all respects in which it is good. In the case of one's own life, therefore, every aspect of value at every moment—one's existence, fine qualities and capacities, enjoyments and achievements, beneficial connections with natural and social environments, and so on—is to be attributed to God's agency as creator. While there is a good deal of disagreement within Christianity on this matter, Christians commonly affirm that God is also behind the bad things that happen, at least insofar as those bad things can be turned to good account (for example, harm suffered turned into a salutary pedagogical correction, just punishment for sin, the necessary testing of one's faith, or simply a beneficial form of sympathy with God's own suffering on the cross). For both the general reasons just mentioned—because of its holism and temporal inclusiveness—belief in God as the creator of the world encourages love, gratitude, and trust in God, and toward the world that God brings about, as constant Christian dispositions and basic Christian attitudes of wide-ranging applicability, whatever might be going on in one's life.

Social connectedness and invisibility

The same all-inclusive causal dependence upon God at all times ensures that individuals are never left on their own, never abandoned to their own devices. Christian theologians (especially in the Protestant tradition) usually develop the psychological implications of this in terms of avoiding either anxious or arrogant self-concern.[5] According to this theology, one does not believe one ever operates independently of God. Therefore, one should never attribute successes and achievements in a prideful way to oneself, but rather one should always give the glory to God as their ultimate source. For the same reason, one should never despair of failings, as if one's own inadequacies were the last word; one believes a supremely powerful and loving influence, God, remains an operative force in one's life, however, desperate the situation appears to be from the standpoint of one's own powers and capacities to improve one's lot in life.

By discouraging isolated self-regard or self-understanding generally, the same nexus of Christian ideas about God as creator has clear consequences for loneliness or perceived social isolation. Because they believe themselves to be creatures of God, Christians feel related to God whatever happens. Whatever their social circumstances—regardless of how isolated or strong their feelings of abandonment by human others—individuals are to remember that they remain in a relationship with God who is concerned about them. Even when they feel themselves utterly forsaken by others, Christians have reason to believe God cannot be forsaking them. They can believe they are never alone even when they appear to be absolutely so. In such circumstances, Christians can always avail themselves of a completely counterfactual sense of social connection with the best-connected "superfriend" of all: the God who remains, they believe, in a relationship of ultimately beneficial causal efficacy with not just themselves, but everyone and everything.

The very unapparent, counterfactual character of God's influence on human lives—the invisibility of God's influence, in short—permits

Christians to perceive their relationship with God as an unbreakable constant. Because God's influence is unapparent or invisible, Christians can continue to believe God is operative for creation's benefit in the absence of any of the obvious confirming evidence required in ordinary cases of beneficent human influence. Christians who believe that God is a universal influence for good as the world's creator do not expect God to be present in the way one expects a human person to be. Therefore, God's apparent absence, in human terms, need not break down their sense of being in relation to God.

Having God as one's creator (in the best-case scenario) is like having a perfectly loving human benefactor, but the unusual invisibility of this benefactor allows Christians to think God present even when not apparently so. Being in a relationship with God for Christians who believe God is a good creator is something like being in a relationship with a human person who never lets one out of her sight and who intends one's good comprehensively. It is, for example, very like being in a relationship with a loving parent who is fully responsible not just for the fact of one's existence, but in a comprehensive way for one's nurturance throughout one's whole life. Unlike a relationship with an ordinary human person of that sort, however, God is believed to be invisible, and this is what allows individuals to assume God's constant presence, despite all appearances to the contrary.

The invisibility that underlies the Christian affirmation of God's constancy here is a function of the very diffuseness of God's influence, a diffuseness of influence that no human person, invisible or otherwise, could possibly match. Belief in God as the creator of the world does not encourage one to single out God as the cause of any specific happening in a way that suggests God is one cause among many, the cause of this particular happening rather than some other with a different cause. Belief in God as the creator of the world does not allow one to identify God's influence in overly close fashion with any particular causal influence of a beneficent sort. Instead, as I have suggested, whatever is of benefit to the individual over the long or short term is to be attributed to God's influence.

The Christian cannot, then, locate or pick out God as one could a loving parent from within the field of variously operative forces or influences on one's life. For that reason, the Christian need neither fear God's loss nor rue God's absence as one would such a parent's. Unlike relations with human others, which are situation sensitive and thereby susceptible to change of character, rupture, and decline, Christians believe that God is with one, whatever happens, in exactly the same capacity—as the creator and sustainer of whatever remains good about one's life, even if only the bare fact that one continues to live. The Christian who believes God is his or her creator is therefore confident that God continues to work for his or her ultimate good and that God is engaged in the effort to increase it, regardless of whatever impediments in human life suggest the contrary—that is, without almost any confirming evidence.

The problem of inattention

Although invisibility and apparent absence do not pose the same problem here as they do when God is one's friend, this rather more impersonal understanding of God's influence as creator and sustainer has its own problem maintaining a strong sense of connection to God. The diffuseness of influence that lies behind God's constant invisible presence can prompt simple Christian inattention to God. The very monotony of the always pertinent Christian affirmation that everything is to be attributed to God can make that affirmation recede from focal awareness. Belief in God's uniform presence would thereby become functionally indistinguishable from the sense of God's absence. The invisibility of God that follows from a belief in the comprehensiveness of God's influence simply means in that case that God drops out of sight and mind—that is, God drops out of Christian consideration most of the time. Such a God has little to offer as a "para-social" entity, as a factor fomenting or supplementing a sense of social connection.

In the back of their minds, Christians may believe that God is the source of everything, but they may not feel compelled to consider that fact actively in the course of their everyday lives. At the center of attention are all the ordinary influences and connections with one's natural and human environments; preoccupation with them pushes out of focal awareness the fact that God is their ultimate source. Apart from specifically religious obligations—say, the demand to give God thanks and praise at times of worship—Christians who believe God is their creator would have no particular reason to dwell on that fact.

Christian theologians commonly tie this sort of practical worry—about what from a Christian point of view amounts to sinful neglect of one's connection to God—to the fact of double agency in the account of creation I sketched earlier.[6] *Double agency* means that both God and created causes act at once, on two different causal planes or levels, to bring about what happens in the world. God acts on the world from outside it to create and sustain in existence the very created causes that act within the world to make things happen. When created causes are sufficient to explain what happens within the world—which is most of the time—one can easily neglect the fact that God must also be acting on the world to create and sustain those causes. God seems obviously necessary to explain events that do not have sufficient created causes—for example, to explain how Jesus is resurrected from the dead. But in the ordinary run of life, where created causes seem sufficient to account for what happens, the need for God's agency as creator and sustainer easily recedes from view.

The temptation to lapse into habitual obliviousness of one's relationship with God is countered, however, by other beliefs about God that Christians hold. For example, the common Christian belief that God acts as more than a creator in individual lives helps prevent obliviousness to God. God does not merely act as creator by giving individuals the created gifts that make them what they are—for example, their own capacities and operations, the ability to influence and be influenced by their human and natural environments, and so on. God also acts to give them his very presence—by way, for example, of their relationship with Christ, who Christians believe was God in human flesh. The very presence of God in human life means one's relationship with God cannot be ignored. The created causes and influences, through which God also influences human life, consequently no longer have the same capacity to distract human attention from God.

Christians often believe, moreover, that God's direction of human life by way of his own presence is no optional matter: God's presence forms an essential component of human life. In addition to created capacities and influences brought about by God, his presence is necessary for ordinary human capacities to operate as they should.[7] To be morally good, for example, requires not just virtuous capacities of one's own, given to one by God, but the presence of the Holy Spirit within one. Knowing well requires not just the formation of good ideas through the usual human processes of investigating one's environment—the entirety of which has its source in a good creator God—but also a mind informed by the very Word of God. And so on.

Such beliefs imply that attention to God's presence, some sort of God-directedness, should be a constant feature of an individual's every-day, ordinary life for that life to be lived well. The individual Christian is accordingly given a reason to bring to mind his or her relationship with God and is motivated to attend to that relationship as much as possible in the effort to lead a better life. An active God-reference becomes part of a prospective, goal-oriented process of self-reformation in accord with what is believed to be God's intentions for one.

Beliefs like this about God's presence as a constituent feature of human operations are at times incorporated by Christians within an account of creation: God's presence within them is then believed to be an element of what God as the creator of the world gives to every human being and is, in that sense, part of the natural or ordinary constitution of human life that God intends in creating the world.[8] But more often than not, the gift of God's presence as an effective influence on human life is specifically associated by Christians with salvation. Christians typically believe that humans have either lost altogether or, at a minimum, habit-ually fail to attend properly to a presence of God that is always theirs, in ways that corrupt human well-being. The Christian claim is that God saves human beings by giving them the presence of God as an effective force for human transformation by virtue of something that Jesus suffers or accomplishes.

God acts as an invisible force in human lives here because God influ-ences humans through his very presence. Christians who follow the com-mon teaching of theologians in this regard believe that God is invisible or unapparent because God is not capable of being delimited or circum-scribed by the usual boundary drawing and sorting mechanisms used to cordon off and pin down other things.[9] God is not, in short, a kind of thing, set off by clear boundaries that distinguish God from what He is not. But there is also here the kind of invisibility discussed earlier: the invisibility of apparent absence in human terms.

Christian claims about salvation often have an eschatological edge. That is, they frequently point to an end time, indefinitely deferred from the perspective of anything achievable in this life. What God gives to remedy the sin of human life through Christ is, accordingly, not commonly thought to be fully effective in any visible way in this life. Christians typically think that their connection to Jesus brings with it a new availability of the presence of God as a force for change in their lives, but what they expect to achieve by way of that constantly available relationship remains invisible in the form of an always deferred hope. Once again, invisibility—here, the invisibility of the revolutionary changes in one's life for which one continues to hope—permits Christians to believe that the presence of God, made available to them in a new way in Christ for the very purpose of bringing about those changes, is nevertheless always with them.

Conclusion

The main intention of this chapter was to make the case that Christian beliefs about the way God influences human life are likely to have a bearing on perceived social isolation. After suggesting that Christian beliefs of this sort are diverse, I developed a particular construal of Christian beliefs about God's influence that seems to hold great potential to alleviate perceived social isolation by directing attention to one's connection with God. While this argument was merely a logical or prima facie one, it forms a testable (though, as yet, untested) hypothesis: Do people actually feel less lonely when they hold such beliefs? Can they be made to feel less lonely by calling them to mind? More specifically, how does the influence of this construal on feelings of social isolation compare with that of other construals, such as a construal that directly associates God's creative influence with the irredeemable bad? How might the stronger sense of God's presence in the hardships one suffers balance out in the latter case against the unhappy quality of the connection? Might one feel oneself to be better off alone, in other words, if God is as much one's tormentor as one's benefactor? Finally, comparable problems to the ones for belief surface in the more experience-driven God-as-friend outlook in Christianity and make experimental testing pertinent.[10] If the problem in both cases is that a strong sense of connection with God is hard to sustain—because God is invisible in one case and crowded out by more obviously pressing matters in the other—how is the imaginative force of the idea of relationship with God better shored up? By imagining that one is on a date with God, or by imagining that God is always all around one like the air one breathes or the sun that shines? And what works for the greater number of people? What if the former imaginative capacities are hard to cultivate and require exceptional abilities of concentration or inward focus that many people lack? Might beliefs be easier for most people to hold in mind without sustained or disciplined practice? Would a simple visit to church or occasional perusal of a prayer book do?

Endnotes

1. For the general importance of belief for such investigation, see Chapter 11, "Anthropomorphism: human connection to a universal society," by Clark Gilpin; and Chapter 12, "How does God become real?" by Tanya Luhrmann.

2. See Chapter 12.

3. Thomas Aquinas is one prominent theologian in the Christian tradition who highlights this idea: "[C]reation … is the very dependency of the created act of being upon the principle [God] from which it is produced." See *Summa Contra Gentiles, Book Two: Creation*, trans. James F. Anderson (London: University of Notre Dame Press, 1975), specifically Chapter 18, Section 2, p. 55.

4. See again, for example, Aquinas, *Summa Contra Gentiles, Book Two: Creation*, Chapters 31–38, p. 91–115.

5. See, for example, Martin Luther, *Lectures on Galatians* (1535), in *Luther's Works*, Jaroslav Pelikan, ed. (St. Louis, Mo.: Concordia Publishing House, 1964), Vols. 26–27.

6. Aquinas again provides a clear theological exposition of this view. See, for example, his *Summa Contra Gentiles, Book Three: Providence, Part 1*, trans. Vernon J. Bourke (London: University of Notre Dame Press, 1975), specifically Chapter 70, Section 8, p. 237: "It is also apparent that the same effect is not attributed to a natural cause and to divine power in such a way that it is partly done by God, and partly by the natural agent; rather, it is wholly done by both."

7. For prominent examples of such a view in the history of Christian thought, see Augustine, *The Literal Meaning of Genesis*, trans. John Hammond Taylor (New York: Newman Press, 1982), specifically Chapters 10–12, p. 49–51; and Cyril of Alexandria, *Commentary on the Gospel According to John*, trans. P. E. Pusey (Oxford: James Parker, 1874), specifically Book 1, Chapters 7–9, p. 66–87.

8. See, for example, Athanasius, "On the Incarnation of the Word," trans. Archibald Robertson, in Philip Schaff and Henry Wace, eds., *Nicene and Post-Nicene Fathers*, Vol. IV, Second Series (Grand Rapids, Mich.: Eerdmans, 1957), specifically Sections 3–5, p. 37-39.

9. See, for example, Gregory of Nyssa, "Answer to Eunomius' Second Book," trans. M. Day, in Philip Schaff and Henry Wace, eds., *Nicene and Post-Nicene Fathers*, Vol. V, Second Series (Peabody, Mass.: Hendrickson, 1994), specifically p. 257.

10. See Chapter 12.

The elusiveness of meaningful connection

Kathryn Tanner developed the classical Christian idea that God is the creator and sustainer of the world, to suggest the ways in which this notion of the creator might be one factor providing persons with a sense of social connection and a hopeful, generous, and caring disposition toward the world that assuages the adverse health consequences of loneliness. In this classic interpretation of God as creator, the idea refers not to the origins of the universe, but rather to the all-inclusive dependence of life upon God at all times. This sense of a sustaining divine presence spanning the whole time of one's life thus contributes a deep sense of one's connection to the whole order of creation. However, as Tanner notes, people may become inattentive to a presence so pervasive, just as people can become inattentive to the forces of gravity holding them to the surface of the Earth as they go about their everyday life. In more extreme versions of this inattention, the person understands humanity as "alone" in the universe, a sort of metaphysical loneliness that might exacerbate more concrete feelings of loneliness.

Perhaps surprisingly, Chris Masi, from the perspective of a physician and medical researcher, casts a fascinating and fresh perspective on the theological notion that we live in a sustaining connection to creation as a whole. After describing the negative health consequences of loneliness, Masi proceeds to describe a cycle of loneliness in which a person's sense of isolation frustrates well-intended efforts to make social connections. Masi finds that efforts to intervene and break this cycle are not notably successful, in large part because the preconscious disposition of lonely people toward the world is difficult to change. Like Tanner, although using different terminology, Masi's review of the scientific literature suggests both that the character of one's general disposition toward the world is profoundly important for one's connections to others, and that the processes by which these general dispositions change are complex and warrant further scientific attention.

14 *

Visible efforts to change
invisible connections

Seventeenth-century English philosopher Thomas Hobbes proposed that, without the organizing structure of government, humans would experience *bellum omnimum contra omnes* (war of all against all) and life would be "solitary, poor, nasty, brutish, and short."[1] While this colorful description is often quoted, less attention is paid to Hobbes's premise that such misery can be avoided if humans codify and enforce the rules of a civil society. Not everyone agrees with Hobbes's views, but history is

---

* The lead author is Christopher M. Masi, M.D., Ph.D., an Assistant Professor in the Department of Medicine at the University of Chicago. He is the cofounder of Every Block A Village Online, an Internet-based community development program, and is past president of the Illinois chapter of Physicians for a National Health Program. He is the current president of the Midwest Society of General Internal Medicine and has received numerous awards, including a Models That Work Award from the United States Bureau of Primary Health Care and the

replete with examples of human misery when anarchy reigns and of relative peace when a social contract is observed. A question that philosophers continue to debate is whether collaboration for mutual benefit is part of human nature or whether promotion of the self above all others is man's primary motivation. In Chapter 2, "The social nature of humankind," John Cacioppo argues that sociality is an integral part of human nature. He notes that, given each child's prolonged period of total dependence, survival into child-bearing age depends entirely upon the support and protection of adults, most often parents or kin. As a result, those who survive long enough to procreate pass along genes for nurturing and protection, thereby hardwiring a form of sociality into our genetic code.

This protective behavior helps ensure that genes within a family are passed on to future generations. Cooperation among unrelated adults or the support and protection of children by adults exists beyond kin as well because these activities also provide survival benefit. Examples from early human existence include hunting and gathering, which are more likely to succeed when pursued as a group than individually. To these structural benefits of nonfamilial sociality, we may now add physiological benefits. A 1979 population-based study showed that adults lacking social ties were 1.9 to 3.1 times more likely to die during a nine-year follow-up than those who had more social contacts, all else being equal.[2] Since then, at least five population-based prospective studies[3] and

New Investigator Health Sciences Research Award from the Gerontological Society of America. Masi has a medical degree and a Ph.D. in social service administration. His research focuses on the socioeconomic factors underlying health disparities. He currently has two projects, one aimed at developing an intervention to reduce loneliness and one focused on the role of sex hormones in gender, age, and racial differences in cardiovascular disease. He is a reviewer for several scientific journals and grant-making organizations and has published research and reviews on diverse topics, including health insurance reform and racial disparities in breast cancer and hypertension.

Human capacity for creativity, compassion, and learning is unparalleled in the animal kingdom. However, humans reach their full potential only when they are socially engaged. Lack of social engagement impairs creativity and learning, and limits opportunities for caregiving and emotional growth. Numerous studies have shown that loneliness is also a risk factor for illness and premature mortality. Because loneliness is increasing in modern society, it is critically important to understand this condition and strategies to reduce it. This chapter describes our review of the literature regarding loneliness reduction interventions.

numerous smaller studies have found positive associations between social integration and either survival or improved health outcomes. The mechanisms by which social integration enhances survival are several and include improved health behaviors, increased access to resources and material goods, and strengthened immunity against infection.[4]

Whereas sociality is a normal state, loneliness is an unusual state, akin to hunger, thirst, or pain.[5] As with those states, loneliness is unpleasant and serves to remind us that we should change the *status quo*. Therefore, loneliness can be an adaptive motivator for increased social surveillance and interaction. Unfortunately, not all individuals succeed at achieving the level of social connectedness they desire and suffer instead from chronic feelings of loneliness. Cacioppo and others have shown that lonely individuals interpret events and social interactions more negatively than nonlonely individuals. As a result, they unconsciously develop defense mechanisms, including social barriers, which shield them from insults and rejection. While this approach may achieve its goals of self-protection, it also reduces opportunities for positive social interactions and perpetuates feelings of social isolation.[5] John Bunyan, a seventeenth-century Christian writer and preacher, described the barriers associated with loneliness when he wrote of a vision in which he

> saw the people set on the sunny side of some high mountain, there refreshing themselves with the pleasant beams of the sun, while I was shivering and shrinking in the cold, afflicted with frost, snow, and dark clouds. Methought, also, betwixt me and them, I saw a wall that did compass about this mountain.[6]

For chronically lonely people, the wall between themselves and others is partly of their own making and reflects continuous surveillance for negative signals from others.[5] The challenge is to help lonely individuals break down the barriers between themselves and others and ultimately return to the normal state of sociality. In his vision, Bunyan achieved this, but only through great effort:

> About this wall I thought myself, to go again and again, still prying as I went, to see if I could find some way or passage, by which I might enter therein, but none could I find for some time. At the last, I saw, as it were, a narrow gap, like a little doorway in the wall, through which I attempted to pass; but the passage being very strait [sic] and narrow, I made many

efforts to get in, but all in vain, even until I was well-nigh quite beat out, by striving to get in; at last, with great striving, methought I at first did get in my head, and after that, by a sidling striving, my shoulders, and my whole body; then I was exceeding glad, and went and sat down in the midst of them, and so was comforted with the light and heat of their sun.[6]

Genetic studies indicate that heritability accounts for approximately 50% of loneliness, while social circumstances account for the other 50%.[7] Research also suggests that loneliness is common, reported by as much as 20% of the population at any given time.[8] In addition, some evidence suggests that the prevalence of loneliness may be increasing, at least in the U.S. A recent national survey found a threefold increase in the number of Americans who indicated they had no confidant or person with whom to discuss important matters.[9] Although differences between this survey and its 1985 predecessor may be sufficient to account for this increase, this suggestive report raises the possibility that contemporary societal factors may be interfering with the natural tendency for humans to form meaningful long-term social connections. One factor is social mobility, which increased dramatically during the twentieth century. A second is the aging of the U.S. population. In 1900, 4.1% of Americans were 65 years or older. By 2006, that percentage had increased to 12.4%, representing 37.3 million Americans.[10] With less value placed on older individuals in the U.S., we have witnessed an increase in marginalization of this segment of society. Third, as life expectancy increases, more elders are living longer as widows or widowers and are therefore at increased risk for loneliness. Other factors that may place Americans at increased risk for loneliness include less intergenerational living, delayed marriage, increased dual-career families, increased single-residence households, and increased age-related disabilities and health conditions. Given the mental and health risks associated with loneliness described in Louise Hawkley's Chapter 4, "Health by connection: from social brains to resilient bodies," interventions are needed to help lonely individuals regain normal social connections. As Bunyan's account suggests, breaking through the wall of loneliness may require considerable effort. When individual effort is not sufficient, assistance from others may be needed. Unfortunately, contemporary interventions to reduce loneliness have fared more poorly than has been recognized.

Repairing broken connections

Almost a century ago, scholars began to propose strategies for reducing loneliness. Karen Rook,[11] for instance, amassed over 40 interventions dating back to the 1930s in her attempt to identify effective loneliness-reduction strategies. Since Rook's review, five scientific publications have provided qualitative reviews of strategies to reduce loneliness, social isolation, or both.[12-16] The most recent publication identified 30 interventions published between 1970 and 2002,[16] and evaluated the effectiveness of those intervention studies that were not flawed by poor design. Among the 13 trials deemed to be of high quality, 6 were considered effective, 1 was considered partially effective, 5 were considered ineffective, and 1 was inconclusive. The authors' conclusions were similar to those of prior reviewers, who found that interventions that emphasized social skills training or opportunities for social interactions were the most successful. Other intervention strategies, including enhancing social support and altering maladaptive social cognition, appeared less successful.

However, qualitative reviews are subject to invisible biases that can color our judgments of the scientific evidence we see. Thomas Kuhn, a twentieth-century physicist and epistemologist, noted that scientists too easily accept results that conform to previous intuitions and too readily reject results that do not.[17] In the case of loneliness interventions, all of the reviews essentially confirmed the findings of previous reviews that social skills training and group-based interventions can succeed in reducing loneliness. Is this conclusion justified, or is this a case in which prior conclusions have been perpetuated in the manner Kuhn describes?

To combat bias favoring results that confirm dominant theories, some scientists have argued that specific study criteria should be met to warrant an evaluation of the effectiveness of the intervention.[18] These criteria include random assignment of study participants to receive the intervention, evidence that the intervention is more effective than no intervention, findings that are replicated by at least one independent research group, and results that are published in peer-reviewed journals. Previous reviewers of loneliness interventions have, in fact, placed a premium on randomized trials that contrast a group randomly selected to receive the intervention with a group randomly selected to receive no intervention. However, none has employed meta-analysis, a quantitative technique for calculating the average effect of diverse interventions designed to accomplish the same goal. Whereas qualitative reviews are subjective and vulnerable to confirmatory biases, quantitative reviews are objective and relatively impervious to bias as long as all relevant studies are included in the analysis.

To minimize bias in our meta-analysis, we first combed the literature to identify all the intervention studies that specifically targeted loneliness. To further meet our criteria for inclusion in the meta-analysis, the studies had to be published in a peer-reviewed journal or as a doctoral dissertation (to ensure the scientific integrity of the findings) between 1970 and 2009 (to include and extend the time interval reviewed qualitatively in prior research) and had to measure loneliness quantitatively.

Fifty intervention studies for loneliness met our inclusion criteria. These studies were divided into three categories based on the experimental design used to assess the effects of the intervention. Twelve studies used a single group pre-post design in which loneliness among participants was assessed at baseline and again after exposure to the intervention. The single pre-post design is weak in terms of measuring the effectiveness of an intervention, however, because individuals who have high scores on a loneliness measure on one occasion are likely to score less extremely on a second occasion even if no intervention had occurred. Said differently, people whose measurements suggest they are very lonely at one point in time, on average, appear to be less lonely when measured at a later point in time. Our meta-analysis of these studies indicated there was indeed a lowering of loneliness as measured before and after the interventions, but we cannot conclude from this evidence that the reductions in loneliness were due to the interventions.

Eighteen studies utilized a nonrandomized group comparison design in which some of the participants sought out the intervention (the experimental group), while others (the control group) did not. In this design, assignment of individuals to the experimental or control groups was based upon convenience, participant preference, or some other factor, which means the groups that did and did not receive the intervention may differ in ways that explain different outcomes in the two groups. For example, people who volunteer to be in the experimental arm of a study may be more gregarious by nature and may be more likely to become less lonely over time *regardless* of exposure to the intervention. Results of the meta-analysis suggested the interventions might be effective, but to know it's the intervention and not an artifact of subject selection, we need to look at the effect of these interventions when assessed using randomized group comparison designs in which participants are randomly assigned to the experimental or control group.

Twenty studies utilized such a design. Quantitative analysis revealed that, on average, the interventions had a small but significant effect in reducing loneliness. Moreover, interventions that addressed maladaptive social cognition were more successful than those that emphasized social skills training, increased opportunities for social interaction, or enhanced social support. Whereas once a consensus existed that social skills training and/or increased social interaction could reduce loneliness, we found that changing the way lonely people thought about themselves and others yielded the best results. Unfortunately, only four of the twenty studies utilized this approach.

Why have successful interventions to reduce loneliness been so elusive? Several possible reasons exist. Some of the interventions have been designed with the notion that if only lonely individuals had better social skills, they would be able to form satisfying connections with others. However, recent research suggests that, at least for most adults, the social skills they know are not related to the loneliness they feel. Other interventions have been developed with the notion that lonely individuals simply need to interact more with others, so the interventions are designed to increase contact with others. However, people not only tend to like lonely individuals less than nonlonely individuals, but lonely individuals are especially negative toward other lonely individuals. Therefore, bringing lonely individuals together is unlikely to result in warm, satisfying social connections. Finally, some interventions were designed with the notion that what lonely individuals need is social support, such as someone who is available to provide help when needed. However, loneliness affects not only how people think, but also how people think about others: Loneliness diminishes people's executive functioning and biases them to see others as threatening and rejecting even when they are not.[5]

Cacioppo and Patrick[5] proposed a framework for reducing loneliness that builds upon research documenting the centrality of maladaptive social cognition to loneliness. First, unconscious barriers that chronically lonely people develop to shield against being hurt by others tend to reduce their likelihood of having positive social interactions; therefore, they may benefit from encouragement and practice in forming social connections gradually in "safe" environments where threat of rejection is minimal. For instance, because chronically lonely people are self-focused in their hypervigilance for social threat, they may benefit from learning to shift their attention from themselves to others through other-oriented activities such as volunteerism. The notion is to intervene to diminish or eliminate the negative effects loneliness can have on social perception and cognition. Second, we tend to think of loneliness as the same thing as a personal weakness, as being a social isolate, being depressed, or being weak. As noted earlier, we now know that these accounts are incorrect and that acute loneliness, just like acute physical pain, serves an important biological function for our species. Being aware of how loneliness fits into our remarkable achievements as a social species and what loneliness does to our social cognition and behavior can help us better understand the actions of others toward us. Third, to the extent that desperation for social connections leads chronically lonely individuals to misguidedly vest their interest in those who are unlikely to meet their relationship needs, they may need to learn how to be selective and choose friends and groups with whom reciprocally rewarding relationships can be expected. This decision is critical to success. Research indicates that the people with whom we are most likely to form positive, lasting relationships are those who have similar attitudes, beliefs, values, interests, and activities to our own. Therefore, people should not seek friendships based on physical appearance, status, popularity, or convenience, but rather on attitudes, beliefs, values, and behaviors. Finally, because chronically lonely people expect to be disappointed with themselves and others in their relationships, they may benefit from training and practice in taking a more optimistic perspective, in expecting the best from themselves and from others. We play a much more important role in shaping our social environment than we often realize.

Although no intervention to date has incorporated all of these elements, at least one randomized trial has demonstrated that an intervention based upon volunteerism (Experience Corps) can increase social activity in older adults.[19] In this trial, older adults were paired with grade-school children and dedicated at least 15 hours per week throughout the school year assisting the teachers in supporting and encouraging children in reading, writing, and mathematics. This strategy engages at least two of the principles that emerged from Cacioppo and Patrick's theoretical framework[5]—the provision of a "safe" venue for making social connections (such as the classroom of nonthreatening children), and the shifting of older adults' attention away from their own concerns and toward the needs of someone else. In addition, this strategy capitalizes on Erikson's notion of generativity (such as helping future generations).[20] Interventions of this form deserve further assessment.[21]

Conclusion

We began this chapter by noting that loneliness is not uncommon and, although unpleasant, may prompt individuals to attend to and repair their social connections. Loneliness affects cognition as well as well-being, however, and when loneliness persists, it is a risk factor for myriad health problems. Previous reviewers have suggested that loneliness can be reduced through interventions that emphasize social skills development and increased opportunities for social interaction. After quantitatively analyzing twenty well-designed studies, we found that these strategies were less successful than those that modified maladaptive social cognitions. This approach should be replicated and enhanced using the elements described by Cacioppo and Patrick.[5] In the interim, it is clear from this review that global impressions and intuitions will not suffice when trying to reduce loneliness. Future interventions should acknowledge that loneliness is not synonymous with social isolation but is a social pain that functions to motivate the formation and renewal of meaningful social relationships. When feelings of loneliness fail to accomplish their adaptive purpose, chronic loneliness may ensue. Chronic loneliness tends to be self-perpetuating through confirmatory biases that alter cognitions, emotions, and behaviors. Given the importance of social connection to people's health and well-being, it is important that we solve the puzzle of how to help the chronically lonely connect with others in meaningful and satisfying ways.

Endnotes

1. T. Hobbes T, *Leviathan, or the Matter, Forme, and Power of a Commonwealth, Ecclesiasticall and Civil* (1651). Available at www.earlymoderntexts.com/f-hobbes.html.

2. L. F. Berkman and S. L. Syme, "Social Networks, Host Resistance and Mortality: A Nine-Year Follow-up Study of Alameda County Residents," *American Journal of Epidemiology* 109 (1979): 186–204.

3. J. S. House, K. R. Landis, and D. Umberson, "Social Relationships and Health," *Science* 241 (1988): 540–545.

4. L. F. Berkman and T. Glass, "Social Integration, Social Networks, Social Support, and Health," in L. F. Berkman and I. Kawachi, eds., *Social Epidemiology* (New York: Oxford University Press, 2000).

5. J. T. Cacioppo, W. Patrick, *Loneliness: Human Nature and the Need for Social Connection* (New York: W.W. Norton & Company, 2008).

6. J. Bunyan, *Grace Abounding to the Chief of Sinners* (Grand Rapids, Mich.: Baker Book House, 1986).

7. D. I. Boomsma, G. Willemsen, C. V. Dolan, L. C. Hawkley, and J. T. Cacioppo, "Genetic and Environmental Contributions to Loneliness in Adults: The Netherlands Twin Register Study," *Behavioral Genetics* 36, no. 6 (2005): 745–752.

8. D. E. Steffick, "Documentation on Affective Functioning Measures in the Health and Retirement Study," University of Michigan Report No.: DR-005 (2000).

9. M. McPherson, L. Smith-Lovin, and M. E. Brashears, "Social Isolation in America: Changes in Core Discussion Networks over Two Decades," *American Sociological Review* 71 (June 2006): 353–375.

10. Administration on Aging, *A Statistical Profile of Older Americans Aged 65+,*: Department of Health and Human Services, 2008.

11. K. S. Rook, "Promoting Social Bonds: Strategies for Helping the Lonely and Socially Isolated," *American Psychologist* 39, no. 12 (1984): 1389–1407.

12. B. T. McWhirter, "Loneliness: A Review of Current Literature, with Implications for Counseling and Research," *Journal of Counseling & Development* 68 (1990): 417–422.

13. M. Cattan and M. White, "Developing Evidence Based Health Promotion for Older People: A Systematic Review and Survey of Health Promotion Interventions Targeting Social Isolation and Loneliness among Older People," *Internet Journal of Health Promotion* (1998): 13. http://rhpeo.net/ijhp-articles/1998/13/index.htm.

14. R. A. Findlay, "Interventions to Reduce Social Isolation among Older People: Where Is the Evidence?", *Ageing & Society* 23, no. 5 (2003): 647–658.

15. E. F. Perese and M. Wolf, "Combating Loneliness among Persons with Severe Mental Illness: Social Network Interventions' Characteristics, Effectiveness, and Applicability," Issues in Mental Health Nursing 26 (2005): 591–609.

16. M. Cattan, M. White, J. Bond, and A. Learmouth, "Preventing Social Isolation and Loneliness among Older People: A Systematic Review of Health Promotion Interventions," *Ageing & Society* 25, 2005: (41–67).

17. T. S. Kuhn, *The Structure of Scientific Revolutions,* (Chicago: University of Chicago Press, 1962).

18. D. L. Chambless and S. D. Hollon, "Defining Empirically Supported Therapies," *Journal of Consulting and Clinical Psychology* 66 (1998): 7–18.

19. L. P. Fried, M. C. Carlson, and M. Freedman, et al., "A Social Model for Health Promotion for an Aging Population: Initial Evidence on the Experience Corps Model." *Journal of Urban Health* 81, no. 1 (2004): 64–78.

20. T. A. Glass, M. Freedman, M. C. Carlson, et. al. "Experience Corps: Design of an Intergenerational Program to Boost Social Capital and Promote the Health of an Aging Society," *Journal of Urban Health* 81, no. 1 (2004); 94–105.

21. J. W. Rowe and R. Kahn, "Experience Corps: Commentary," *Journal of Urban Health* 81, no. 1 (2004): 61–63.

Reflections on invisible connections

Echoing a prominent theme in this volume, Christopher Masi highlights once again the centrality of social connectedness for human well-being and the function of loneliness in signaling a rupture in a sense of social connectedness. One might reasonably expect that a social species like *Homo sapiens* would have a sufficiently large behavioral repertoire to be able to resolve feelings of isolation and restore a sense of social connectedness. Although resolution is accomplished readily in some instances for some people some of the time, sometimes people are at a chronic loss for how to satisfy their need for social connection. Unfortunately, the invisible bonds of social connection are not easily repaired. We see others' social activity, but we do not see how they feel about their social lives and sense of connection. Despite our inability to recognize loneliness in others, or, as Nick Epley and Jean Decety argued earlier in this book, because of this handicap in seeing into the minds of others, we tend to attribute to others what we ourselves have felt or would expect to feel in particular circumstances. Is it any surprise that we target for intervention those circumstances where we observe few opportunities for social interaction, inadequate social skills, and poor social support? On the other hand, because loneliness is in the mind of the sufferer, it is perhaps surprising that we would expect changes in objective social circumstances to be sufficient to alleviate loneliness in all its sufferers. Masi provides a quantitative review of strategies employed to alleviate loneliness to show that interventions to date have been only modestly successful in reducing feelings of loneliness, attesting to the challenge of effectively addressing the problem of ruptured social connections. Nonetheless, a few studies suggest that addressing the root cause of loneliness, namely maladaptive social cognition, holds promise as an intervention strategy.

Invisibility should not thwart attempts to alleviate distress, however. Biological causes of disease were no more visible or evident in the eighteenth century than psychological causes are today. Yet significant scientific advances during the nineteenth and twentieth centuries completely revolutionized medical practice, life expectancy, and quality of life. Farr Curlin is less interested in the invisible causes of disease than in the primordial need for social connection that John Cacioppo introduced and that Curlin regards as a preexistent condition for medicine. If science can be viewed as a cognitive system that steps us back so that we can deal more objectively and effectively with another person's distress, then religion can be viewed as a cognitive system that steps us forward to connect and care for others. Curlin argues that the practice of medicine requires a balance of these forces and that the resulting tension between the two produces better care for the patient than does the practice of medicine using either alone.

15 *

Social brain, spiritual medicine?

No one ever asks what *science* has to do with medicine any more than they ask what books have to do with education or what tools have to do with carpentry. Before the middle of the nineteenth century, there was almost nothing that physicians, however well intended, could do to actually restore health to the ill. Modern science changed that. Over the past century and a half, dramatic improvements in health outcomes have

* The lead author is Farr A. Curlin, M.D., a hospice and palliative care physician, researcher, and medical ethicist at the University of Chicago. His empirical research charts the influence of physicians' moral traditions and commitments, both religious and secular, on physicians' clinical practices. As an ethicist, he addresses questions regarding whether and in what ways physicians' religious commitments ought to shape their clinical practices in our plural democracy. Curlin and colleagues have authored numerous manuscripts published in the medicine and bioethics literatures, including a *New England Journal of Medicine*

been wrought through the application of sterile surgery techniques, specialized hospital care, public health measures to prevent the spread of infectious diseases, antibiotics to treat those diseases, and myriad subsequent technologies. All of these have been undergirded by the discoveries of biomedical science.

As a result, the life expectancy in developed nations has doubled. People live not only longer, but with much less disability. Diseases that formerly disfigured and killed, such as smallpox and polio, have been almost completely eradicated. Epidemics of malaria, yellow fever, measles, and diphtheria have been restrained. Injuries from war or other traumatic events, which in earlier periods led predictably to death or profound disability, now can be ameliorated using sophisticated surgical reconstruction techniques, advanced prostheses, and intensive rehabilitation. Medical science already has accomplished an extraordinary amount in alleviating human illness and forestalling death, and there is good reason to expect further progress. Yet for all that science has made possible, medicine is animated by other, less tangible forces.

To give a robust account for the practice of medicine, one must explain why sick and debilitated strangers are worthy of attention and care, and how the medical arts contribute to human flourishing. For some Americans, such accounts begin in secular moral tradition, but for most, they begin in religion; nine out of ten Americans endorse a religious affiliation.[1] Either way, medicine looks beyond science to find a

paper titled, "Religion, Conscience, and Controversial Clinical Practices." As founding Director of the Program on Medicine and Religion at the University of Chicago, Dr. Curlin is working with colleagues from the Pritzker School of Medicine and the University of Chicago Divinity School to foster inquiry into and public discourse regarding the intersections of religion and the practice of medicine.

In the world of contemporary medicine, science is front and center, and for good reason. Science provides modern medicine with extraordinary diagnostic and therapeutic capacities that can be employed to care for patients. Yet there is more to medicine than science can know. Science cannot provide visions to animate care of the sick, moral frameworks to guide the application of medical technology, or practices that nurture and extend our sociobiological capacity to care for others. For these, medicine turns to religious and secular moral traditions and practices. This essay examines how religious concepts are implicit and operative in practices of medicine and in the formation of fully human physicians. By attending to these concepts, we may gain a richer understanding of the way self-conscious human practices like medicine both depend on and extend our unique, human, biopsychosocial capacities.

vision that animates care of the sick, a moral framework that guides the application of medical technology, and practices that nurture and extend the human capacity to care for patients as persons rather than as mere objects. In this sense, even though religious concepts are rarely made explicit in public and professional discourse about medicine, they are everywhere implicit and operative, and necessarily so.

Why care for the sick?

Humans in all cultures are moved to care for the sick. The question is, why? The concept of the social brain provides the beginning of an answer. The peculiar human need and capacity for constructive, complex and meaningful relationships seems to involve neurological structures and functions that also facilitate attending to the sick. For example, in Chapter 10, "Seeing invisible minds," Nick Epley describes the human capacity to pay attention to our own mindedness and the mindedness of others. Not only are we conscious of ourselves, but we are conscious of others being conscious of *them*selves and of us. This capacity allows us to be mindful of others' bodily suffering and mindful of their consciousness of our relation to them in that suffering. To mindfulness is added the capacity to empathize. Jean Decety describes a neurological structure through which the sight of pain in another person triggers a response in our own brains that mirrors (albeit at a level attenuated by training and other contextual factors) the response we would have if we were suffering the pain ourselves. These features of the human brain allow us to pay attention to and, to some extent, share in the suffering of others—capacities that are psychological building blocks for caring for the sick.

Yet to explain medicine strictly on the basis of empirical science, one must solve a particularly thorny version of the more general problem of explaining altruistic human behavior. Decety notes, "The emergence of altruism, of empathizing with and caring for those who are not kin, is…not easily explained within the framework of neo-Darwinian theories of natural selection." Indeed, one can scarcely imagine a practice less conducive to the reproductive fitness of a population than spending enormous resources caring for the sick, the deformed, the weak, and the aged. Natural selection and the physician would seem to be at cross-purposes: One works to eliminate the sickly, the other to save them from elimination. On this account, medicine appears to be the sort of dead end into which the evolutionary process sometimes blindly drifts.

John Cacioppo, however, argues that altruistic behaviors can be explained within evolutionary theory by paying attention to inclusive fitness and the multiple levels of selective pressure:

> [F]or species born to a period of utter dependency [for example, humans], the genes that find their way into the gene pool are not defined solely or even mostly by likelihood that an organism will reproduce but by the likelihood that the offspring of the parent will live long enough to reproduce…one consequence is that selfish genes evolved through individual-level selection processes to promote social preferences and group processes, including reciprocal social behaviors, that can extend beyond kin relationships.

The concept of inclusive fitness helps to explain why humans care for the young when they are sick, and even why they care for those who, when healthy, are able to contribute to caring for the young. In addition, hunter-gatherers may have been more likely to survive and reproduce when they cared for a wounded or sickened member of the clan, thereby establishing an expectation of reciprocity that would contribute to social cohesion, collective effort, and defense of other group members. These provide at least the rudiments of an evolutionary rationale for the practice of medicine.

Yet medicine does not involve caring merely (or even primarily) for the young, much less for those who are most genetically fit. Rather, medicine in large measure involves caring for either those who have no capacity to contribute to the gene pool because they are aged and otherwise infertile, or those whose contributions to that pool will reduce population fitness because they are genetically predisposed to sickness and disability. Concern about the latter led Francis Galton and many of his American and European contemporaries to embrace social Darwinism and to champion efforts to keep the diminished and infirm from reproducing. In the United States, the eugenics movement was memorialized in the infamous words of Supreme Court justice Oliver Wendell Holmes, who justified the constitutionality of the forced sterilization of mentally "unfit" women in the case of *Buck* v. *Bell* by writing, "Three generations of imbeciles are enough." Sterilization rates under eugenic laws in the

United States increased following this ruling until the *Skinner* v. *Oklahoma* case in 1942, after which point they declined.

The practice of medicine expresses more than a straightforward social instinct for protecting the young. To borrow from Don Browning, it may be that medicine builds on and extends the dynamic of inclusive fitness, much like in Catholic moral theology *caritas* (love) builds on and extends *eros* (desire). Browning writes, "[Aquinas] held—and Christianity has always taught—that Christian love includes more than kin altruism and the care of our familial offspring; it must include the love of neighbor, stranger, and enemy, even to the point of self-sacrifice." The theological concept of God as creator and Father of all "made it possible for Christians to build on yet analogically generalize their kin altruism to all children of God, even those beyond the immediate family, their own children, and their own kin." Even those beyond the reasonable hope of reproducing or helping others to reproduce.

Notably, the self-conscious commitments that animate medicine do not include promoting population fitness or ensuring survival of offspring to the point of reproduction. Rather, physicians discipline themselves to practices that make possible the commitment of medicine: to preserve and restore the health of patients, notwithstanding patients' other characteristics. Religions ground this care for the sick in sacred and transcendent obligations to God and neighbor, and it is not incidental that the hospital began when Christian monastic communities enfolded the care of the sick into a communal life of liturgy and prayer. This is not to say that the substantively irreligious lack proper motivation to practice medicine. It is to say that an animating vision for medicine as a good and worthy activity seems to require moral concepts that science alone does not provide.

How should medical science be deployed?

Medicine is not only animated by something like a religious vision; it also requires a thick moral framework for its ongoing *direction*. To know how best to care for patients, we need to know something about what human flourishing entails and how medicine can contribute to it. Medical science is less helpful here than one might hope.

Science facilitates the sort of religious humanism that Browning encourages because it helps us better understand the empirical world and, therefore, helps all moral communities refine their efforts to bring about human flourishing. Science elucidates a range of technical possibilities and provides information about what we can reasonably expect as the consequence of choosing one course over another. Yet even the successes of medical science highlight its limits. As medical science generates technologies that can be put to ever wider uses, it exposes disagreements about which of those uses are worthwhile. Although medicine proceeds in scientific ways in the care of patients, it does so in pursuit of goals that science cannot set. These goals come from moral traditions and cultures, religious or otherwise.

In the same way that the influence of a dominant culture on medical practice is often invisible or taken for granted precisely because of its dominance, the influence of religious ideas on medical practice is often invisible in those areas where commitments are shared among different religions and other moral traditions. For example, we generally take it for granted that mending injuries, treating infections, and removing diseased organs are good things to do. That is because the moral commitments that undergird these practices are shared by virtually all moral communities, religious or otherwise. Moral commitments that are shared by all may not seem "moral" at all. Yet even the idea of sickness implies a norm of and concern for health that are not fully derivable from empirical science.

The influence of religion on medical practice becomes more visible where the commitments of particular traditions diverge from one another or from the values of the dominant culture. For example, religious measures have been found consistently to strongly predict physicians' attitudes regarding ethically controversial practices such as abortion, physician-assisted suicide, withdrawal of life-sustaining therapies, contraception, and physician interaction with patients about spiritual concerns, as well as, we have found, physicians' ideas about the relationship between religion and health.[2]

Yet overtly controversial issues merely highlight the tips of proverbial icebergs. Disputes about practices such as abortion and physician-assisted suicide concern whether the practices are intrinsically unethical. Much more commonly, physicians agree about the range of legitimate clinical strategies, but they disagree about which is to be recommended in a given moment. For example, physicians may agree that the experience of depression can be treated legitimately by antidepressant medications, referral to a psychiatrist, or referral to a counselor whose practice is rooted in a specific religious tradition. Yet our research suggests that the religious characteristics of physicians strongly influence which of these options they would recommend in a given case.[3]

Controversies over a particular medical intervention often represent deeper unspoken disagreements that, unfortunately, science cannot settle. For example, controversies over the use of stimulants to manage childhood behavior disorders, or the medicalization of social anxiety seem to reflect disagreements about more basic questions: What brings human happiness? Which moods and behaviors should be considered normal parts of human experience, and which should be considered abnormal? What sorts of suffering should we try to alleviate? What leads to disordered behaviors? What resources (social, psychological, spiritual, or otherwise) are best suited to addressing disruptions in individuals' mental and emotional states? How does modern medicine fit into our response to these experiences? Although physicians may not ask or answer these questions explicitly, they implicitly answer them in their responses and recommendations to patients.

So for all that is hoped for in "scientific" and "evidence-based" practice, clinicians must ultimately act as practical moral philosophers, making judgments about how best to pursue the goals of medicine for a particular patient in a particular context. Among those things to be considered are moral valuations about which religions and other moral traditions have much to say, but about which medical science remains silent.

Caring for the patient as a person

So far, I have suggested that religions provide a vision that animates care of the sick and a moral framework that guides the application of medical technology. Religions make another contribution by fostering practices that nurture the human capacity to care for patients as persons rather than as mere objects.

Patients commonly complain that their physicians treat them as mere objects or specimens instead of appreciating and attending to them as unique persons. This problem has always plagued the profession. To learn how to heal, the novice physician must learn of patients as representing abstract general types and classes. She must learn about coronary artery disease and hematuria before she can begin to interpret Mrs. Smith's chest discomfort and Mr. Jones's red urine. These abstractions allow knowledge of when and how things happen, and that knowledge guides technological interventions that may bring healing to the body. These abstractions also help doctors objectify their patients' humanity enough to violate social norms that operate in every other social situation, such as asking patients to expose their nakedness in vulnerable positions, or cutting patients apart in hopes of making them whole.

As long as the process does not go too far, scientific detachment serves to make our concern effective. Yet the collective experience of both patients and physicians suggests that such detachment usually does go too far and occurs too easily. As a result, physicians treat patients as mere objects and instances of disease; they treat patients as less than the human persons they are.

Physicians, it would seem, are subject to a particular form of the more general psychological challenge of paying attention to other minds. Like all humans, physicians easily ignore the mindfulness of others. This matters, Epley reminds us, "because mindful agents become moral agents worthy of care and compassion." As such, patients who are seen as mindful "evoke empathy and concern for well-being, whereas agents without mindful experience can be treated simply as mindless objects." There are obstacles to recognizing the mindfulness of patients. Illness makes a patient different, or deviant, from human norms, and we tend to pay less attention to the minds of those who are different from ourselves. In addition, "Considering other minds requires some attentional effort. It does not come automatically." Physicians learn to go through the technical motions of caring for the sick until those motions become "automatic"—that is the mark of a skilled and effective clinician. But paying attention to the mindfulness of patients requires a sustained investment of time and energy that physicians are often unwilling to make.

How could religious practices help? As Luhrmann notes, most people find it very difficult to pay attention to God. To help in this difficult and lifelong task, many religions have developed disciplines of prayer and other practices that call to mind what we tend to forget—including the ideas that motivate genuine human concern for those who suffer. Christians, for example, practice remembering that all people are ultimately united as children of the one creator God—that "the ground is level at the foot of the cross," regardless of one's social status, biological fitness, or reproductive capacity. Epley notes that we are better able to pay attention to what another is thinking or feeling when we are motivated to do so. Christianity seeks to stimulate such motivation by encouraging Christians to meditate on the fact that Jesus comes to us in those who are sick and otherwise suffer.[4] Moreover, it reminds us that we are never alone. As Kathryn Tanner details in Chapter 13, "Theological perspectives on God as an invisible force," God is always with us. This central theological claim, when remembered in song, prayer, liturgy, Scripture, and other rituals, provides a particular form of what psychologists call "mindful surveillance": Our actions become more "prosocial" (even altruistic) when we are aware of being observed by others. All of these practices depend on and extend the capacities of the social brain. They are also, from the vantage of Christianity, ways in which one may come to receive grace, the unmerited help of God.

Religious practices thus have at least the potential to encourage and strengthen the human capacity for attending to the mindfulness—and, therefore, the personhood—of those who are sick and diminished. As Epley suggests, "Making minds visible, and hence more like one's own, enables people to more readily follow the most famous of all ethical dictates—to treat others as you would have others treat you."

Conclusion

Science and religion are invisibly and inextricably intertwined in the practice of medicine. Science has provided modern medicine with extraordinary diagnostic and therapeutic capacities that can be employed to care for patients. Science gives knowledge of the remarkable neurological and psychological features of the social brain that make activities like caring for the sick possible. But science can also depersonalize the patient viewed through the eyes of the physician scientist. Religions (and other moral communities) motivate an attention to the person who is the patient, providing a fuller vision for the worthiness of caring for the sick and drawing the physician and patient closer together. Religion and moral communities can also provide a framework to guide the application of medical science in that endeavor and practices that strengthen the human capacity for treating patients as the mindful persons they are. The balance of the tensions produced by the forces of science and religion may hold a key to better medical practice and patient care.

Endnotes

1. F. A. Curlin, J. D. Lantos, C. J. Roach, S. A. Sellergren, and M. H. Chin, "Religious Characteristics of U.S. Physicians: A National Survey," *Journal of General Internal Medicine* 20, no. 7 (July 2005): 629–634.

2. See F. A. Curlin, M. H. Chin, S. A. Sellergren, C. J. Roach, and J. D. Lantos JD, "The Association of Physicians' Religious Characteristics with Their Attitudes and Self-Reported Behaviors Regarding Religion and Spirituality in the Clinical Encounter," *Medical Care* 44 (2006): 446–453; and F. A. Curlin, S. A. Sellergren, J. D. Lantos, and M. H. Chin, "Physicians" Observations and Interpretations of the Influence of Religion and Spirituality on Health," *Archives of Internal Medicine* 167, no. 7 (2007): 649–654.

3. F. A. Curlin, S. Odell, R. E. Lawrence, M. H. Chin, J. D. Lantos, K. G. Meador, and H. G. Koenig, "The Relationship between Psychiatry and Religion among U.S. Physicians," *Psychiatric Services* 58, no. 9 (2007): 1193–1198.

4. Holy Bible, Matthew 25:40.

Invisible forces

Farr Curlin meditates on the puzzle of medicine—what its evolutionary and social function is, what draws individual practitioners to it, and what grounds its fundamental values. The values of scientific inquiry lead to treating the objects of inquiries in just that way: as objects. But objectifying patients and their disease seems to work against the human values of empathy and caring for the weak that also seem to be part and parcel of what medicine is as a practice. Curlin argues that religious values inform and nurture the human side by insisting that there must be a connection between physician and patient, acting as an often unrecognized invisible force that humanizes the practice of medicine.

Religion is neither necessary nor sufficient for an individual to adhere to such values. The question of what grounds the fundamental values that govern our relationships and how those values are reflected in invisible social, psychological, and biological forces is central to the work of our network. In Chapter 16, "Epilogue," Ronald Thisted reflects on the many threads of investigation and discussion that have made up our conversation, and how they are interwoven into a network of inquiry that sheds light on invisible forces and the social brain.

16 *

Epilogue

Over the past six years, our network of scholars has engaged in an ongoing conversation that we have come to recognize as being centered on unseen forces that shape, and are shaped by, the social nature of human beings. The chapters that make up this book hint to what our conversation has been like, but the linear structure that a book imposes cannot

---

* The lead author is Ronald Thisted, Ph.D., a professor in the Departments of Health Studies, Statistics, and Anesthesia & Critical Care at the University of Chicago, where he currently chairs the Department of Health Studies. Trained in philosophy and mathematics at Pomona College and in statistics at Stanford University, he maintains interests in the nature of argument and evidence, particularly in the context of health, disease, and medical treatment. He has published articles on topics ranging from treatment for back pain to computational mathematics, and from social determinants of health to the size of Shakespeare's

fully evoke the give-and-take of vigorous debate, the excitement of viewing an old problem from a new perspective, or the satisfaction that comes from sharing the search for knowledge—even when we did not agree on the interpretation of what we discovered in our search.

We deliberately chose to describe our membership as a network rather than a committee, seminar, task force, club, or salon. A network is defined as much by the connections between people as it is by the individual people themselves. Networks can be described pictorially as nodes (points that represent individuals), some of which are connected by edges (lines that represent links between two individuals). In our network, we have focused on the value of the edges and have held the conviction that much is to be gained by exploring previously untested connections. We started with a set of nodes that had only a handful of edges, and we ended with many more edges than nodes.

As a result, our network—and each individual in the network—has been enriched as we have learned more about, and from, perspectives that initially were unfamiliar to each of us. The end result was that our whole is decidedly greater than the sum of our parts. This illustrates a recurrent theme in the book, of emergent phenomena—characteristics that can be ascribed to entities at a higher level of organization that, without conscious design or intent, seem to arise from behaviors and interactions at a lower level of organization. How this can come about is a puzzle, but it is a puzzle that is amenable to thoughtful investigation, both scientific and philosophical. What forces are at play, we might ask, to make such a collection cohesive? Just what chemistry can transform a collection of individuals into something both more than and different

vocabulary. He is a Fellow of the American Statistical Association and a Fellow of the American Academy for the Advancement of Science.

The question of how we come to know—or to claim that we know—that a proposition is true is all too often left unexamined. The similarities and differences in modes of argument across disciplines, and the variations in what counts for evidence supporting or refuting a position within and across disciplines, can be illuminating. Statistics, and statistical argument, provide a rich framework for thinking about such issues as measurement, learning, uncertainty, variation, and experiment. Statistical principles provide a framework for disciplined investigation, for communication about the extent of and limitations to the information at hand, and for combination of information from different sources. Although there is enormous variability among individuals, commonalities to their experience transcend their differences. As a species and as individuals, we rely on these common threads, even when they are invisible to us.

from what in aggregate they bring to the table? We seek to understand more fully the bonds of marriage, family, friendship, or membership—invisible forces that bind and simultaneously transform the underlying nature of their constituents no less than chemical bonds transform atoms of hydrogen and oxygen into water.

Our origin was rooted in distaste for the unproductive and unenlightening shouting matches between those who propound views of science that denigrate religious belief, and those whose views of religion are anti-scientific. We started from the assumption that scholars both from the sciences and from religion and philosophy could have fruitful conversations about what is known, what counts for knowledge, what can be observed, and what can be tested through experiment and observation. And we all believe in the value of the scientific method as a means for expanding our knowledge. Internal tension is needed for the structural integrity of buildings and bridges, and that is no less true of social structures such as our network. Through appropriate construction, deep tensions between theology and science (or even between scientific disciplines or theological perspectives) that have the potential to drive us apart can instead be shaped to release creative energy and shared purpose.

Gary Berntson notes that "beliefs and emotions have consequences, both behavioral and physiological." The network starts from the premise that one can learn about such apparently invisible phenomena as beliefs by studying and reasoning about their consequences.

In his Chapter 3, "Science, religion, and a revived religious humanism," Don Browning advocates starting with a critical hermeneutic phenomenology, a "careful description" of our instruments, our observations, and the stories we use them to tell. Clearly articulating our assumptions and starting points has been of immense value. After doing so for the benefit of colleagues outside our disciplines, those colleagues have helped us become aware of unarticulated assumptions implicit in our approaches or in our experiments. These observations have led to better science and more convincing evidence. Our colleagues in the network have helped each of us to see more facets of the same elephant that individually we are too blind to appreciate fully.

Revising our thinking and our research to take those observations into account has increased the rigor of our thought and broadened the scope of our conclusions. The presence of a rich variety of disciplinary perspectives has helped us weave the nets of Sir Arthur Eddington's parable more tightly, enabling us to see for the first time some of the "smaller fish" that earlier would have escaped our notice.

Shedding light on invisible forces (a koan)

Invisible forces of culture, connection, and curiosity bind us together and define us as a species that is at once both individual and social. Because both individuality and sociality are fundamental to the human species, we are fundamentally interdependent, connected by invisible yet powerful threads. In exploring these threads, we have also been led to questions about how social forces can have effects on individuals, how the meaning that individuals (and groups) apply to particular phenomena or relationships affect both behavior and biology, and how our biology makes social connection possible. We have used the phrase "invisible forces" to describe the mechanisms that account for these effects that we essentially take for granted, and to suggest by analogy that they can be investigated rigorously just as other phenomena, such as gravity or autonomic regulation, that also are not immediately present to our visual or other senses can be studied.

Human minds are unparalleled at discerning patterns in what they see against a background of noise and variation, and they are equally adept at attributing meaning to them. As the chapters in this book demonstrate repeatedly, we readily ascribe patterns we encounter (or seek to encounter) to invisible forces of nature, God, kinship, genes, culture, love, and social connection. A common premise underlying the work of the network is that what we know (or think we know) and how we come to know it are social endeavors embedded in a shared view of both the world and how one talks meaningfully about the world. And mindful of our human facility to see patterns (even where none exist), we are acutely aware that constant rigorous testing of assumptions, methods, and arguments is necessary to make sure we are not fooling ourselves into seeing only what we hope to see.

Humans have a deep need to create meaning in their interactions with the world and with each other. We also have a deep need for making connections beyond ourselves. The biological structure we call our brain has evolved to reward social connection, just as it rewards the satisfaction of hunger or thirst. The human biology that directs and reflects these human needs is what we have termed the "social brain."

It is worthwhile to reflect on the range of invisible forces we have considered here. These forces operate at several different levels, from the molecular, to individual bodily functions, to social groups, to societies, to species. They include such disparate ideas as evolutionary selective pressure favoring social connectedness, anthropomorphism, loneliness, social connection, emergent phenomena, connection to a higher being, transcendence, empathy, language as carrier for meaning, belief, collective will, group synchrony, autonomic regulation, and neural resonance. These forces interact with one another, too: Loneliness, for instance, acts as an internal signal of the inadequacy of one's bonds of social connection, with consequent effects on health, mediated through autonomic regulation, or the role of belief in mediating scientific objectivity and empathy.

It is tempting to view individuals, both souls and bodies, as arising from lower-level forces within, such as the operation of specialized neurons and regulatory biological processes. And it is tempting to view social structures and the forces that tend to maintain them as arising, perhaps emergently, from the individuals that make up societies. On this view, the social level of organization arises out of the interaction of lower-level entities. But invisible forces operate in both directions; one's degree of social integration or isolation (at the higher level) can have profound influence on one's mental and physical health (at the lower level). Just how these forces operate—in both directions—is one of the main themes of this book.

A recurring theme is the human need for connection. As we have explored this fundamental need, it has become clear that it can be satisfied in part by connections not necessarily to other human persons, but to other *minds*. Since the minds of others are in part of our own construction, connections to a higher being, to our pets, or even to a transcendent order underlying the world can fulfill part of what we strive to attain. Indeed, such nonhuman attachments can share the character of human connection: We can feel valued by our pet (just as we can feel validated in a social relationship), we can have an intimate dyadic relationship with God (just as we can be intimate with a close friend), and we can feel a sense of belonging to the universe (just as we can feel that we belong to social group). This explains how different, even contradictory, notions of a relationship with God, for instance, can lead different people to find meaning in such a relationship: finding God on the downtown bus versus encountering God in the purposeful unfolding of the natural order.

The ideas of symmetry, complementarity, coordination, and coregulation also run through several of our essays. Regulation of biological systems is often maintained through paired systems of biological checks and balances; when one system is activated, the other tends to restore equilibrium. For instance, one set of muscles flexes the arm, and an opposing set extends it. We have seen that the sympathetic and parasympathetic components of the autonomic nervous system—the system that makes us breathe and that makes our heart pump—operate in this way, and that chronic stimulation of some systems, like overstretched elastic bands, causes them to lose their ability to spring back. The notions of observing a behavior and performing that behavior not only are conceptually similar, but they also may be rooted in a common set of neurological structures that may help us understand how we can perceive another human being as being *like* us, but *not* us. Anthropomorphism is the belief that other minds mirror our own; this colors the way we perceive the world and the other actors in it, a mechanism that allows us to simulate getting under the skin of the other person.

Happiness and loneliness are perceptions about our place in the world that profoundly affect our physical bodies *and* our social relationships. Religious beliefs, too, can have profound effects on health and physical well-being, working through the same biological mechanisms that, in health, maintain equilibrium.

Unseen yet powerful forces regulate social behavior. Empathy, for instance, contributes to the regulation of social interactions. Synchronous behavior points to a phenomenon that makes the individual feel subsumed by the group, feeling part of a larger, organic whole. These behaviors can be as disparate as "the wave" at a sports stadium or congregational prayer at a church service. Shared feelings of transcendence and belonging can simultaneously lead to greater fitness of the individual and increased cohesion and sustainability of the social organization—another indication of positive selection associated with the social brain.

The notion of resonance with another appears repeatedly through the book. Our connections to others derive in part from being able to see what they see, to hear what they hear, to know what they know, and to feel what they feel. We have to be able to believe not only that this is possible, but that it happens. The social brain, in which the same regions are activated by our own experience of pain and by our perception of others in pain, makes both aspects possible. There is a close connection between being able to "feel for" another (empathy) and to "see into" another mind (anthropomorphism).

Language has the potential to affect people and groups in part because it is tied to meaning. Language is the medium through which we convey, preserve, and transmit meaning from one individual to another and from one social generation to another. Language is powerful because it can activate belief, which can activate physical responses. Words can bind, words can terrify, and words can cause physical pain and death. The power of words comes from the meanings they entail about our connections to one another.

Paradox

Our investigation of invisible forces involving the social brain has led us repeatedly to factors that fundamentally conflict. An important invisible force is the respect we pay to the boundary between self and other. Our relationship to it comes into play in conceptualizing loneliness, anthropomorphism, spirituality, group behavior, empathy, and inclusive fitness. When we speak of loneliness, this boundary seems to be an impenetrable barrier. When we speak of empathy or anthropomorphism, however, the self-other boundary is defined by the similarity and congruence of individuals to one another, providing a transparent window through which we perceive and interact with others (who must be like us). And when we speak of group synchrony, the boundary vanishes completely: Self and other are one.

Successful engagement with others requires work. It is the work of *attending to* something, and it is work that often is needed to resolve competing forces. Thinking about other minds is a demanding task and requires attentional effort. This effort allows us to manipulate the transparency of the self-other boundary by what we put in through learning, attending, seeking, and projecting. In effect, we can tune the degree of resonance we have with members of different groups. Similarly, consistent attentional effort is also required for the physician to attend to the mindfulness of patients, for the Vineyard church member to experience

God as present in one's life, and for another to find connection to an omnipresent yet invisible God who works through the very workings of the world.

What it means to feel a connection to a higher being is a theme that several chapters explore. As is evident in these essays, the Network has considered very different, even divergent, pictures of what such connection might entail. The apparent inconsistencies in these portrayals are rooted in the different aspects of human connections, and each is grounded in a social context. Social connection can be intimate, relational, or collective. For the member of the Vineyard Church, connection with God is an intimate two-way relationship, while in Jonathan Edwards's sermon in the Great Awakening, the connection is relational and involves the coherence (or lack of it) of the individual with God's approval. The Christian theological view of connection as a higher order can be conceived in terms of one's belonging within a whole that God's constancy makes larger than oneself.

While religion certainly speaks to individual connection to others and to the divine, religious practices can also serve an evolutionary and social function by strengthening the human capacity for attending to the personhood of those who are sick and diminished. The objectivity of medical science all too often leads to an objectification of the patient or, more frequently, the patient's disease. The social brain's capacity to see others as minds rather than objects makes it possible to assign meaning to patients and the ways in which they lack wholeness.

Crescat scientia; vita excolatur (where knowledge increases, life is enriched)

The possibility that religion and science can enrich one other, even as one sets aside truth claims about such matters as the existence of a deity, is by no means obvious. But we have come to see that science can describe what religion does in rigorous ways that benefit religion, and religion can serve a meaning-making function that science itself disclaims. Gilpin notes that rifts between science and religion "have centered on whether one can make scientific sense of the notion of divine mind, purpose, or intention." Our network sidestepped this question from the beginning, focusing instead on related matters such as the *consequences* of believing in such a mind, and of seeing into that mind, for the one doing the divining.

Those are questions amenable to empirical investigation, and it is at that juncture that we can see benefit from our discussions. As Berntson says, "Beliefs color the way we perceive the world, they direct and shape our actions, and define our personalities." Studying and debating about how they do so has been gratifying and immensely enjoyable. We have engaged in no theological debate, but have focused on questions about human beings, their beliefs, their behaviors, and how those affect and are affected by multiple levels of human connection.

How we conceptualize our relationships to persons and things outside our selves has implications for our health and well-being. Specifically, we have seen that viewing our relationships in terms of meaningful connections with other minds can have positive implications for individual—as well as social—health and function. The more we can learn about those implications, the more our increase in knowledge has the potential to enrich human life.

About the authors

The **Chicago Social Brain Network** members represented in this book are John Cacioppo (Network Director) from the University of Chicago, Gary Berntson from Ohio State University, Don Browning from the University of Chicago, Farr Curlin from the University of Chicago Medical Center, Jean Decety from the University of Chicago, Nick Epley from the University of Chicago Booth School, Clark Gilpin from the University of Chicago, Louise Hawkley from the University of Chicago, Tanya Luhrmann from Stanford University, Chris Masi from the University of Chicago Medical Center, Howard Nusbaum from the University of Chicago, Gün Semin from the University of Utrecht, Steve Small from the University of Chicago Medical Center, Kathryn Tanner from the University of Chicago, and Ron Thisted from the University of Chicago Medical Center. The biography of each, along with an explanation for the essay each presents, is provided at the beginning of each chapter.

Index

parental investment
  Christian love and, 41-44
  moral implications of, 40-41
patients, as people not objects,
  191-193
pattern discernment, 200
Paul (apostle), 37, 39
penguins, group characteristics
  example, 21-22
perceived social isolation
  dangers of, 26-29
  physical effects of, 28
perceptions
  human boundary with
    nonhuman entities, 136-137
  of others' pain, 114-117
  understanding, 95-97
periaqueductal gray, 115
personal beliefs about God,
  impersonal beliefs versus,
  157-159
personal self, 27
personification of God, 144
Phaedrus (Plato), 85
phenomenology, 37
philosophy, Christianity and, 37
*The Philosophy of Physical
  Science* (Eddington), 29
philosophy of science, causality
  in, 12
physical effects of social
  isolation, 28
physiological effects of
  loneliness, 53-55
pituitary gland, 53
Plato, 43, 85

Plotinus, 38
Politics (Aristotle), 42
Pope, Stephen, 44
prayer, 152-155
preaching, language in, 85
predisease pathways
  health behaviors, 51
  stress, 52-53
presence, fundamental
  expectations of, 147
presence of God in human life,
  165. *See also* reality of God
processing language, 87
projecting onto other minds, 127
psychophysiology, 69

Q-R

randomized group comparison
  experiment design, 177
reading minds, 123-124
  attending to, 125-126
  learning, 124-125
  projecting onto other
    minds, 127
  seeking other minds, 126
reality of God, experiencing,
  145-155
reducing loneliness, strategies
  for, 175-179
regulatory component of
  empathy, 112-114
relational connection/
  isolation, 28
relational self, 27
relationship, experiencing God
  in, 150